Bioeconomy of Buildings

"This book by Jan Grossarth has several strengths. First, it avoids the common pitfall of suggesting that the future of construction should rely solely on wood and biological materials, thus avoiding a simplistic 'return to nature' approach. Instead, it provides a comprehensive overview of the diverse and often conflicting sustainability concepts and ideologies, and presents a variety of promising approaches emerging in the field of construction."

—Prof. Dr. Michael Braungart, *Author of "Cradle to Cradle"*[1]

[1] Translated from the Foreword of the 2024 German edition.

Jan Grossarth

Bioeconomy of Buildings

From Resource Flows to Meanings

Springer

Jan Grossarth
Department of Architecture and Energy
Engineering
Biberach University for Applied Sciences
Biberach, Baden-Württemberg, Germany

ISBN 978-3-031-84013-5 ISBN 978-3-031-84014-2 (eBook)
https://doi.org/10.1007/978-3-031-84014-2

The original submitted manuscript has been translated into English. The translation was done using artificial intelligence. A subsequent revision was performed by the author(s) to further refine the work and to ensure that the translation is appropriate concerning content and scientific correctness. It may, however, read stylistically different from a conventional translation.

This Springer imprint is published by the registered company Springer Nature Switzerland AG
The registered company address is: Gewerbestrasse 11, 6330 Cham, Switzerland

Acknowledgments

To my academic teachers Gunther Hirschfelder and Stephan Schleissing. I also extend my gratitude to my colleagues at *Biberach University* and its *IfH Institute* for their support through interviews, discussions, and/or organizational assistance, including Maximilian Rimmel, Rainer Weitschies, Gordian Blumenthal, Anette Schober-Knitz, Gerhard Haimerl, and Daniel Spruth. Thanks also to many of my students for their insightful discussions. Special acknowledgment to Julian Schleissing for his work on the German book version and to Nik Leon Bürk for his careful assistance. Mr. Bürk is responsible for the English version's editing—mindfully using available AI tools—and he contributed to the layout of all illustrations (unless otherwise noted). His assistance was funded by *IfH Institute*.

Contents

1 Introduction: Bioeconomy of Buildings—Energy, Resources and Meanings

Abstract

Why discuss sustainable construction in such abstract terms as the 'bioeconomy of buildings'? This introductory chapter focuses on three main reasons for doing so. The first is the specific *emphasis on energy* within the tradition of bioeconomic thought. The second reason is the potential to bring a *historical and intercultural perspective* into focus. Within the framework of bioeconomy, reflecting on (past) agrarian bioeconomies becomes relevant, especially in terms of land allocation and its uses for food, fiber, fuel, and feed over time. Historical examples can serve as 'distant mirrors' helping to reflect on present techno-economic approaches. By contrast, the focus on sustainability is often tied to short-term, entirely quantifiable goals, particularly those needed for certification. Finally, the third reason for discussing a bioeconomy of buildings relates to the term itself, which is sometimes used as an intriguing metaphor: *bio* suggests life, connecting to a cultural dimension of sustainability. By examining the lifecycle and beyond of materials, and the cultural meanings embedded in architectural practices, we can rethink buildings as part of a broader ecological system, where resources are circulated and environmental impact is reduced.

Whoever speaks of the "bioeconomy" does so with critical awareness of the global economy's (fossil) energy dependence. Although there are different traditions surrounding this term, this is their common ground. This critical awareness is a simple consequence of a historically well-informed view of current dependencies. From the economy as a whole to its segments, especially in the fields of architecture and construction, energy dependence is anything but irrelevant. From earth to bricks, from iron to concrete, the history of buildings cannot be understood without considering the history of agriculture and energy supply (Calder, 2021). Energy, forestry, mining or transport revolutions explain the choice of forms and materials,

J. Grossarth, *Bioeconomy of Buildings*, https://doi.org/10.1007/978-3-031-84014-2_1

whether in ancient temples, medieval churches, or the architecture of modernism up to the present time.

In one of its two main traditions, bioeconomic thinking warns about dependency and seeks more reliable paths of long-term development. In the other, it focuses on the potential contributions of increased usage of timber, straw, or other biomass material innovations (see Fig. 1.1). This book addresses both traditions and connects them to the fields of construction and architecture. Readers will gain insight into the history of current problems, the contributions of technical and social innovations, and the legal framework. The book discusses circular, sustainable options in a broad sense, while also exploring the historical development of the concept(s) of bioeconomy.

Buildings are energy- and emission-intensive in both construction and operation, more so than at any time before the 1950s. This book's approach is to connect the ideas, histories, innovations, and politics of the bioeconomy to this field of application. It aims to achieve two practical purposes. First, it outlines pathways (and barriers) toward a significantly more resource-efficient building sector. Second, and somewhat unusually, it highlights the personal, experiential qualities associated with housing—such as joy, memory, or stewardship. Through the concept of bioeconomy, understood not only by official definitions but also as a rich and polysemous metaphor, these qualities might with certain limits be integrated into the sustainability discourse of housing. As the book's subtitle "meaning" suggests, it is also about political relevance. In other words, it explores the function of the bioeconomy concept(s) in discourses, such as the long-running focus on sustainability, and lastly, diverse personal meanings that might be associated with this metaphor. A word like "bioeconomy" raises different associations at a personal level, possibly evoking images of beautiful, characterful, or durable architecture, as the prefix "bios" suggests something vital or life-sustaining (see later chapters).

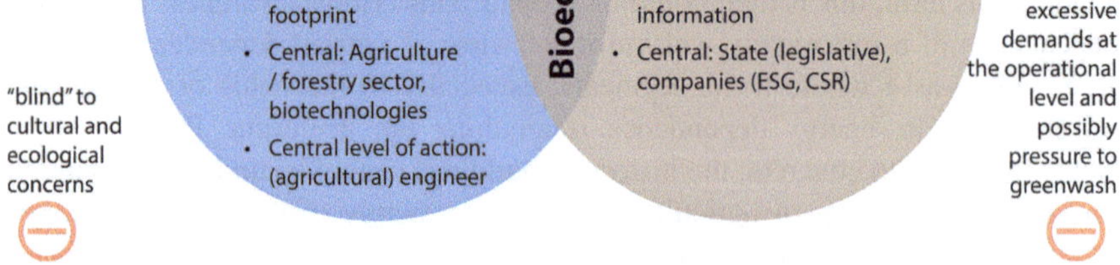

Fig. 1.1 Two traditions of bioeconomic thinking. JG

Bioeconomy of buildings focused on materials and energy in a long run. Over a lifetime, the U.S. population consumes an amount of energy comparable to what would be required to build 50 million replicas of the Pyramid of Khufu, a 4,500-year-old monument constructed from 5 million tonnes of sandstone, limestone, and granite (Calder, 2021, pp. xii–xiii). This nowadays consumption is made possible by construction machinery, global transportation, and raw material exploitation. What might be labeled as "sustainable" or a "green building" must not be viewed as such from the critical perspective of our ongoing extraordinary energy dependence. It is important to clarify that the concept of bioeconomy, while related to the global sustainability focus established in the mid-2020s, is distinct from it. The bioeconomy represents an approach with unique institutional and ideological developments, dating back to the 2000s and even the 1960s. Today, it particularly emphasizes renewable raw materials and more efficient material flows, reflecting its long history influenced by economics and environmental engineering.

Hence, it is important to clarify that the concept of bioeconomy, while related to the sustainability focus established globally in the mid-2020s, is distinct from it. The bioeconomy is an approach with unique institutional and ideological developments dating back to the 2000s and even the 1960s. It particularly nowadays emphasizes renewable raw materials. In the context of architecture and the construction industry, the bioeconomy focuses on 'biomass' (renewable raw materials) and material flows, reflecting its long history influenced by economics and environmental engineering.

Anyhow, as a category, bioeconomy is new to the particular case of the sustainability debate about the construction industry, and vice versa. For the past 30 or so years, research on the bioeconomy has primarily addressed agriculture, energy production (biomass), fuels, and renewable materials for the chemical industry. This book aims to close the gap in the construction sector by integrating literature from civil engineering, architectural sciences, forestry, and resource economics in an interdisciplinary manner. (These aspects will be addressed in Sects. 2.2 and 2.4, Chaps. 4, 5, and 6, Sects. 7.1–7.8.) Some chapters examine how the term "bioeconomy" carries specific meanings and interpretations. These interpretations range from the sceptical views on the material sustainability of economic growth in the 1970s to discussions about the cultural context of of both, bio-based innovation but also human experienced of *joie de vivre*, shaped by the living environment. This part of the book integrates cultural studies and hermeneutics, citing literature from the humanities and social sciences (Sects. 2.3, 3.2, 3.4, 3.6, 7.9–7.12).As both people and the planet face critical environmental, economic, and social challenges, global temperatures are expected to rise by 2–8 °C by 2100 (Scafetta, 2024), and the 2020s have witnessed unprecedented species extinction rates (Frankel, 2024a, 2024b). In response, many governments are turning to a circular bioeconomy to achieve a regenerative and carbon–neutral future (Gardossi et al., 2023a, 2023b; Hodgson et al., 2022a, 2022b; Holden et al., 2023a, 2023b). Central to this approach is improving material and energy efficiency, driven by innovations in the life sciences (Lewandowski, 2018a, 2018b). However, transformation politics often face resistance, particularly from populist movements in the EU and US.

These groups, focusing on social identities and local issues, have slowed political efforts, as seen with the 2024 delays to the EU Green Deal due to farmer protests (Oxford Analytica, 2024). After two decades of focusing on quantitative factors like 'biomass availability,' there is now a growing call to examine individual drivers and barriers to transformation in the mid-2020s (Wojtynia et al., 2023).

Not only political, but also academically, the issue of sustainability has become increasingly complex. Perhaps this is the reason why architect Florian Nagler's research houses in Bavaria have been attracting so much public attention in Europe since 2019. "Building Simply" was the working title of the project. Three houses with flexible floor plans, and each as mono-material as possible. One made of masonry, one made of lightweight concrete, one made of solid wood (Nagler, 2021; Fig. 1.2). Quite simple.

But the question of what is sustainable is less simple, as seen from the perspective(s) of sciences. That is, from life cycle assessment to architecture, from ecology to ethics and historical discourse analysis. Rather, the debate about it can be represented with the famous Indian parable of the blind people standing around an elephant. It goes something like this: One feels the tail, another the knee, another the toes, a fourth the back, a fifth the tongue. One thinks it is a snake, the second thinks it is a bone, the third, he would be dealing with stones, the fourth thinks of a carpet, and the fifth of a dead fish: The parable is tailor-made to describe the academic (and especially the media-political) discourses on sustainability in construction. One sees the life cycle assessment, the second the *ESG ratings*. Others see the concrete to be overcome, and others: too large living space, misleading industry standards, capitalism—or a short-lived architecture that has given its aesthetic soul to the mammon of transience.

But that's not how the concept of sustainability is designed, within whose framework a "bioeconomy of buildings" should be located. It highlights areas of conflict. It does not simplify. Sustainability can be seen as an ethical framework for defining and reflecting dilemmas (Henkel et al., 2023). Is does not mean a "holistic view", but a broad view that does not overlook conflicts of objectives and contradictions. That is short versus long term, economic versus ecological, social versus ecologic. Bioeconomy of buildings seems to be fruitful in that fields, where it helps to overcome conflicts, whether through historical inspiration or technical innovation.

Many examples are given in this book. Bioeconomy offers a strictly transdisciplinary perspective. Resource-saving, circular construction seems unattainable from the world of academic silos alone. It also needs the creative passion of craftsmanship, planning, architecture. For the implementation—the planning and realisation—of sustainable architecture, the decision for *one* of the sustainability approaches might be effective. This can be Nagler's "Building Simply", or another: For example, a material-based approach (a house made of straw, like Werner Schmidt's Swiss made straw ball houses, or a house made of hemp). This can be a circular approach like Urban Mining or Cradle to Cradle.

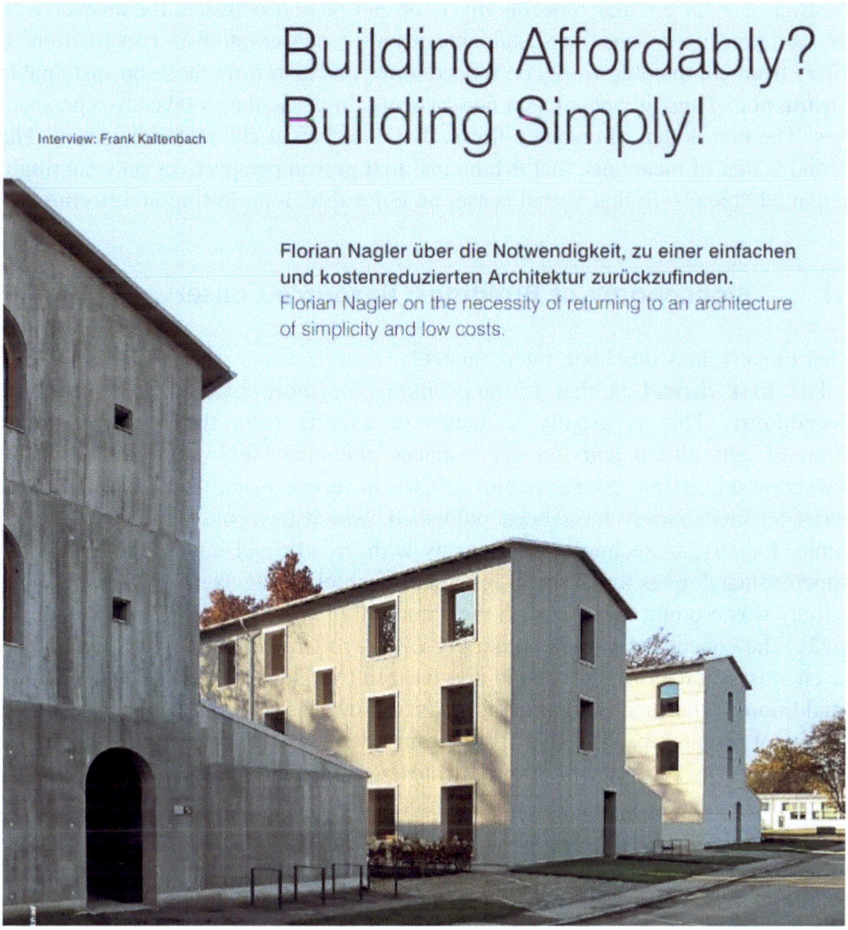

Fig. 1.2 Florian Nagler's "building simply"—an early 2020s metaphor that captures the spirit of the time. Screenshot JG

The academic view does not simplify. Engaging with the discursiveness and also the contradictions of sustainability approaches can nevertheless also be *productive*. The reader might see sustainable building studies as an open minded journey to different *islands of sustainability*, but such a journey does not lead into a world free from dilemmas. Such a travel experience enables one to speak more knowledgeably and reflectively about sustainability. But what can another, ultimately abstract, "academic" or in some respects planning approach like that of bioeconomy bring? Like the circular economy? Should one speak of both separately, or connected—of the circular bioeconomy? Which research is in the individual academic disciplines that deal with bioeconomy and sustainability central? This book provides an overview and seeks answers. It starts by asking the question of what circular bioeconomy is or can be and connects the answer with very different approaches to sustainable resource conservation in construction. It differs from comparable works on bioeconomy, and also from those on sustainable construction. Thus, it wants to set new accents. In particular, it takes two perspectives. The first is that of resource flows, that is technical and material related. The second is that of meanings, that is humans' first person perspectives on what might be named "bios"—in that verbal sense, an enjoyable, long-lasting architecture.

1.1 Bioeconomy of Buildings: Resource Conservation

What threads does this book interconnect?

The **first thread** is that of bioeconomy, or more precisely: the **circular bioeconomy**. This is usually a matter of experts from the academic disciplines of agricultural and forestry sciences and biotechnology, (Kircher, 2020; Lewandowski, 2018a, 2018b; Schurr, 2020). In recent years, significant overview works on bioeconomy have been published, which have often linked the agronomic, forestry, biotechnological aspects with growth and innovation-theoretical "superordinate" questions (Thrän & Moesenfechtel, 2020; Pietzsch, 2017). Bioeconomy was brought together with the paradigm of sustainability (Lanzerath et al., 2022). The concept of circular economy *Cradle to Cradle* was first formulated by the chemist Michael Braungart and popularized (McDonough & Braungart, 2002). In addition to the mentioned monographs and overview works, an overwhelming number of research papers have appeared in recent years. Science journalists tried to bring the concept of the bioeconomy closer to a wider audience. They occasionally came to the conclusion that it could not actually be popularized (Grefe, 2016). It is by good reasons considered too abstract, as a buzzword or "container term". The term is predominantly used in academic and ministerial institutions. However, its ambiguity can be seen as a strength. The bioeconomy is multi-faceted. Clear-cut terms that can be reduced to binary concepts (like sustainability) can also become unconvincing, "burnt out" over time, and have limited topical relevance. The word "bioeconomy" does not suggest a simple solution to the dependency problem. Instead, it invites a range of interpretations and encourages dialectical considerations.

The **second** thread that is to be linked here with that of the circular bioeconomy is that of **low-emission, resource-saving construction**. On this topic, architects such as Felix Heisel or Hans Kollhoff have predominantly expressed themselves, and so did civil engineers like Werner Sobek in an European context. A broader and more radical implementation of such postulates is still pending but is clearly emerging in the 2020s due to climate protection laws and plans in the EU, US or PR of China. Recently, there have also been comprehensive reviews of engineering and environmental science aspects of sustainable construction (Bartels et al., 2022; Bauer et al., 2013; Friedrichsen, 2018). This also applies to approaches and technical potentials of building material recycling (Müller, 2018), circular construction and related digitisation (Heisel & Hebel, 2021; Campanella et al., 2022; Schwarzwälder, 2023).

This book is also a mediator. It brings together research results of many different scientific disciplines what otherwise has little knowledge of each other. For what do structural engineers know about land use competition? What do architects know about *genome editing*? And what do economic and social politicians know about life cycle assessment? On the other hand, what do plant breeders or biotechnologists know about modern laminated timber, and what do real estate agents know about life cycle assessment, and what do building policymakers know about agroecology? But all of that is part of a low carbon, circular bioeconomy of buildings.

The result is intended to be an overview, and may not delight highly specialized scientists with new findings from their field. One concern is the connection of the "puzzle pieces" of research with the long path of economic and energy history. The urgently recommended program of a rapid decarbonisation appears in the long agricultural and industrial historical view as faced with *wicked problems*. That is, taking into account the historically developed path dependencies and mentalities. Also, in view of the rapidly increased world population with rapidly growing consumption needs in the emerging countries, global decarbonisation seems to be challenging.. The vehemence of early 2020s global political climate protests alone did so far not solve the complexly intertwined problems. Probably these are not even "solvable". Not only EU's climate laws have been passed. But the concrete conditions for the implementation of the goal of a net-zero-emission construction industry by 2045 are still partly in the dark.

Given the existing climate protection legislation and partially foreseeable resource shortages, it can be assumed in the mid-2020s that the construction industry is facing a profound and industrial-historically epochal turning point. This book seeks approaches to how this can succeed while preserving the social, economic and ecological interests of the many actors and citizens. It is intended to stimulate new perspectives of research and application. What new perspectives are meant? The specific raw material needs and carbon storage capacities of the construction industry are one. Another is not literally "new", but still largely foreign to the strongly nature and engineering science shaped professional traditions: This is a cultural scientific, hermeneutic view of the topics of bioeconomy and sustainability of buildings. This approach is particularly relevant in the case of the "circular bioeconomy of buildings", as the factor of longevity is particularly important here. The cultural sciences focus on meanings. The connections between the sustainability orientation and the personal meanings people assign to their homes are relevant, particularly because these connections relate to motivations for preservation and, consequently, the longevity of buildings overall.

1.2 Bioeconomy of Buildings as a Metaphor

So, as mentioned above, this book is *not just* about the facts, progress, policies, barriers and definitions of the circular bioeconomy of the construction industry. It also considers narratives about it.

An innovative point of this book is an understanding of the *bioeconomy as a metaphor* (see Sects. 3.6 and 7.12; Fig. 1.3). That simply means, it connects the two abstractions of *bio* and *economy*. With the method of understanding bioeconomy as a metaphor, an attempt is made to bring the many approaches mentioned above together. In other words: it enables a *transdisciplinary discourse*. Thus, observation and interpretation of coexistence—i.e., the parallelism of different *sustainability narratives*—become a concern.

Metaphors matter. They create social and political orientation, majorities organise themselves through them. What does this imply for the *bioeconomy*? Bioeconomy semantically combines biology (if you want to translate it that way: *the living*) with economics (household doctrine). Economics also means the analysis of scarcities. The metaphor is not newly invented, but it remains original. Its political originality lies in the fact that it brings a new dialectic into "dusty" discourses on sustainability. Often contradictory used, opposingly perceived thoughts come together here. The bioeconomy metaphor opens to dialogue and thought. From the economic side, for example, it can be considered to what extent concerns of *naturalness*, but also of life, i.e., *experience-relatedness* of human involvement in economic processes are recognized. From the side of socially resonant strong ideas of sustainability, one can ask whether the conditions of the citizens' and companies' householding are sufficiently "considered" by their representatives.

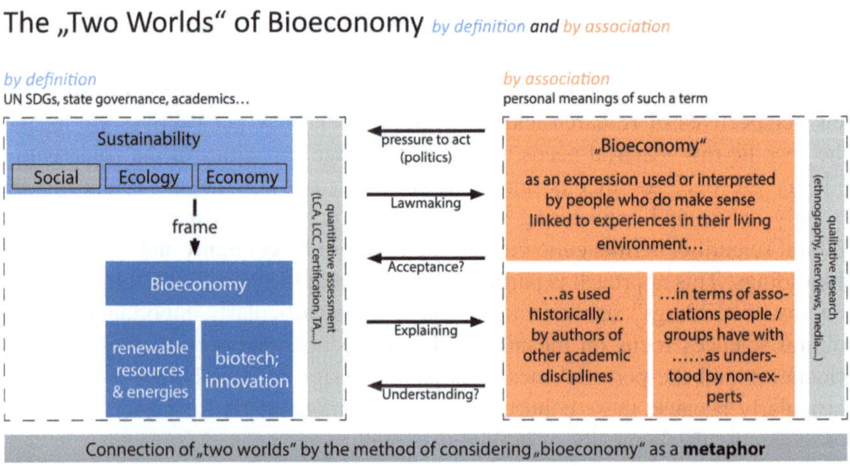

Fig. 1.3 The method to "read bioeconomy" as a conceptual metaphor might link concepts and personal perceptions. JG

The philosopher Hans Blumenberg sees the metaphor in general as something not entirely decipherable; it preserves its "secrets" against all attempts at deconstruction, but also of wanting to understand. The metaphor, writes Blumenberg, "preserves the richness of its origin, which abstraction must deny" (Blumenberg, 2014, p. 90). In this sense, it would be foolish to see bioeconomy simply as a framing term, as an abstract ideal of future economic activity, or also technocratically narrowed.

The metaphor, Blumenberg continues, preserves the foundational connection between "lifeworld and world of theoretical facts" (ibid., p. 94). It is a *bridge of mediation* in this sense—from the "torn" areas of causal-objectivist "thought maps" and the human lifeworld of experiences (ibid., pp. 87–106). Thereby, experiences of life are meant: feelings, the discomfort, the astonishment, the unspeakable, the intention, transcendent experience, and so on. Blumenberg then also sees the metaphor as a "border zone of language, in which writing down would be shame in public, without the claim to have perceived something being withdrawn." (ibid., p. 96).

Thus by metaphorical speaking one can something *perceived* bring to language, without speaking from the first-person perspective, which is irrelevant in the apparatus of science. The metaphor preserves experiential values even under the objectivity constraints that shape the mode of arguing in the science or political discourse. Metaphorical speaking is a poetic border area of unavailability (Müller & Schmieder, 2016, p. 781). And therein may also lie a charm of the metaphor of bioeconomy—namely, because Bios means "life" and might be understand in a broader sense as "vitality" (Fehrenbach, 2011).

However, it should be considered that the bioeconomy metaphor as research policy wording comes from the border area of science and politics itself (Chap. xx). Blumenberg's hermeneutics examined cultural guiding metaphors such as voyages, icebergs or light-filled cave exits. His writings precede the linguistic explanations of the metaphorical structuring of language in general, and the experience-based understanding (Lakoff & Johnson, 2008).

But the bioeconomy also has a second origin, which is particularly not so well known in technical circles. For Nicholas Georgescu-Roegen, who used this term from the 1960s onwards, the bioeconomy denotes a civilisational design, the utopia of a radically ecologically sustainable economy and lifestyle. When reading his long classic and for decades almost even in economic academic circles forgotten texts, it becomes clear, that here also the life experience of an economic scientific author "resonates" (Chap. xx). Roegen has been marginal in his time. His stubborn insistence on the probable decline of energy-hungry "American" forms of civilisation can certainly be reasoned by his personal experiences with dramatic epochal breaks of two world wars. Roegen's involuntary exile in the United States in the face of the communist rule in his Romanian homeland supervenes. Here, the personal standpoint, the "lifeworld experience" cannot be overlooked, which is hidden in the prefix "Bio" (life)—the vague hope for human wisdom, which anticipates and avoids the disasters of resource scarcity. And this function has the bioeconomy metaphor also—with Georgescu-Roegen, earlier also with others:

It has a meaning as a *corrective metaphor*, which one-sided technicist, therefore overbearing scientific conceptions like the infinite economy growth "opposes something".

Hans Blumenberg discusses the metaphor at all in relation to the ambivalent role of science in a technological society (a term by Jacques Ellul; Ellul, 1954). The central role of science as a common intersection of the "reasons" in a modern society has devalued the life experience. "Between the books and reality is an old enmity set." (Blumenberg, 2020 [1986], p. 17). Metaphorical hermeneutics accompanies Blumenberg's tremendous technological progress drive of the post-war decades (Blumenberg, 2015). He writes:

> We no longer know exactly why we undertook the whole vast enterprise of science - independent of all the achievements that it provides for the viability of our world and that make it indispensable for this - at all. (Blumenberg, 2014, p. 93).

Also understood in this way, and applied to our topic, the approach of *bioeconomy* contains more than its biotechnological traditions of genome changes and material flow management, of biomass optimisation and renewable building materials. Understood as a metaphor that has "emerged" historically in different contexts, from Georgescu-Roegen and earlier on, it can "subliminally" express a quite ambiguous, but not blatant momentum of crisis perception and point to the openness of the future and human creativity.

Blumenberg does not play off the areas of abstract knowledge and life experience against each other. He owes much to technology and science (Blumenberg, 2015). The hermeneutics of technology—or: of technology criticism—resigns itself to not capitulating as humans in the hope of making it useful for the joy of life:

> What remains for man? Not the 'clarity' of the given, but that of what he himself has produced: the world of his images and creations, his conjectures and projections, his 'imagination' (Blumenberg, 2013, p. 12).

According to Blumenberg, the metaphor thus refers back from the world of logos and science to the *originally motivating world of life and experience*: Isn't it also an enrichment to think about what it could mean for a house to be like a mammoth, as Michael Braungart suggests in the preface? And: Isn't the desire for the "simple" also one for the assertion of sensual qualities?

> In this sense, metaphors are guide fossils of an archaic layer of the process of theoretical curiosity [...] The mystery of the metaphor cannot be understood solely from the embarrassment about the concept. What is puzzling is why metaphors are tolerated at all. (Blumenberg, 2014, p. 87)

For this book, the necessary external expert knowledge of many scientists and construction practitioners, bio-economists, biotechnologists and technology assessment were not only included through literature research, but also through interviews with civil engineers and architects; they have been recorded in the 2024 German version of this book only: thanks to Armin Grunwald, Rainer Weitschies, Dimitrios Toris, Gerhard Haimerl, Daniel Rubin, Iris Lewandowski, Daniela Thrän, Thomas Auer and many others (Grossarth, 2024). At appropriate places, reports supplement the scientific diction. They are titled "Excursion". In more than ten years in which I have written as a journalist about the topics of the bioeconomy, I was allowed to travel to interesting places and make observations that later occupied me as a university teacher.

An Ecological Metaphor vs. Technocratic Construction?Hans Blumenberg, in other words: The observation, the effect of a place, conversation, a situation, the embodied feeling of the observers, is given (weak) validity via the bridge of the metaphor even where rationalisation is necessary—In ministries, authorities, natural and engineering sciences, life cycle assessment calculation, politics, also environmental and climate policy.

The metaphorical speech comes out, without opposing the obligation of scientific reason, also without opposing the technical or legal "reasons" a subjectivism or aestheticism, without radicalisation to a "romantic culture war", without the arrogance of the protest movements. Wouldn't something be gained if the bioeconomy of buildings were seen as a metaphor in this sense?

These photos were taken on a walk through a village in the German Mittelgebirge (Fig. 1.4). They show archetypes of a common settlement image. How do these new or remodeled buildings affect the viewers? How do they affect you?

Personally, for me it is like this: The charming character of the old village centre stands in stark contrast to the buildings on the peripheral streets. These structures can be described as disturbing, inappropriate, intrusive, and loveless, with additional adjectives such as frivolous, isolated, reckless, unimaginative, and characterless. Doors, railings, steps, staircases, and roofs do not harmonize; instead, there is a hodgepodge of designs focused solely on affordability and functionality. Stone terraces and rubble walls encased in wire mesh dominate the landscape. New settlements blur the distinction between village and fields, with each individual building according to their own preferences. The result is not individuality but uniformity and an unprecedented loss of coherence. Does it have to be this way? Is much of it unnecessarily large? From what perspective should such a judgment be made? Is it purely subjective, or is there an objective basis?

The experience of the walk culminates in this question: Has the construction process, as reflected in these impressions, ever been connected (by those responsible) to the idea of a house as a contribution to the village community, or is it viewed exclusively as a private matter?

So, if asked for an metaphorical interpretation, what can be seen here? Monsters of economisation? The petrification of convenience. An aesthetic capitulation? The capitulation of any willingness to contribute to a *more beautiful* neighborhood? A DIY store architecture?

Each of these metaphors has its strengths and weaknesses and does not necessarily imply that hardware stores or concrete markets should be rejected (unlike

Fig. 1.4 Impressions of newer buildings that have emerged around a historic town centre in a a German rural village. Photos by Jan Grossarth

(g)

Fig. 1.4 (continued)

monsters). Concrete engineers and civil engineers working in solid construction should not feel insulted by them. The intention is to highlight something that appears to have been overlooked in the planning process.

Bio-economy as a metaphor, as this example aims to illustrate, would serve as a semantic counterpoint to such commodified economies, even if these economies are excellently sustainability-certified.

Purpose of This BookIt is *also* to ask about effect and experience of how is built. Sustainability quantification never tells the whole story. The bioeconomic approach allows such questions to be addressed politically, economically, from an engineering perspective, and legally. Otherwise, bioeconomy would only be another lost "buzzword" (Vivien et al., 2019), or a fashionable term—with a limited shelf life -, a catch-all term and a discourse phenomenon, which would have its basis in corresponding research funding.

This book is, first and foremost, about summarizing the approaches of the circular economy and bioeconomy for a construction-related audience. Secondly, it aims to convey the specific concerns of the construction industry to readers in the fields of sustainability and bioeconomy. Given that longevity, resource conservation, and the use of existing stock are particularly important, both quantitative and qualitative methods should be considered for scientific support. The division between architecture—open to such concerns in its self-conception as an applied cultural science—and the technical mindset of civil engineers, in light of sustainability goals, is nothing short of tragic. The circular bioeconomy approach cannot undo this tragedy. But it can create new, delicate connecting threads, considering both, personal experiences and historical knowledge.

References

Bartels, N., et al. (2022). *Anwendung der BIM-Methode im nachhaltigen Bauen*. Application of the BIM method in sustainable building, Wiesbaden.

Bauer, M., et al. (2013). Green building: *Leitfaden für nachhaltiges Bauen*. Guide for sustainable building, Wiesbaden.

Blumenberg, H. (2013 [1960]). *Paradigmen zu einer Metaphorologie*. Paradigms for a Metaphorology, Frankfurt/M.

Blumenberg, H. (2014 [1979]). *Schiffbruch mit Zuschauer*. Shipwreck with Spectator, Frankfurt/M.

Blumenberg, H. (2015). *Schriften zur Technik Writings on Technology,Technology*, Berlin.

Calder, B. (2021). Architecture: From prehistory to climate emergency. London.

Campanella, D., et al. (2022). Echte Materialkreisläufe schaffen—Möglichkeiten und Herausforderungen der Wiederverwendung von BaustoffenCreating Real Material Cycles—Opportunities and Challenges of Reusing Building Materials. In C. Jacob & S. Kukovec (Eds.), *Auf dem Weg zu einer nachhaltigen, effizienten und profitablen Wertschöpfung von Gebäuden: Grundlagen—Neue Technologien, Innovationen und Digitalisierung on the way to a sustainable, efficient and profitable value creation of buildings: Basics—new technologies, innovations and digitalization–Best Practices*, Wiesbaden (pp. 517–535).

Ellul, J. (1954). *The technological society*. Vintage.

Fehrenbach, F. (2011). VitalityLebendigkeit. In Pfisterer, U. (Ed.). Metzler *Lexikon Kunstwissenschaft*. Lexicon Art Science, Stuttgart.

Frankel, C. (2024a). *Extinctions: From dinosaurs to you*. University of Chicago Press.

Frankel, C. (2024). *Extinctions: From dinosaurs to you*. University of Chicago Press.

Friedrichsen, S. (2018). *Nachhaltiges Planen, Bauen und Wohnen: Kriterien für Neubau und Bauen im Bestand*. Sustainable Planning, Building and Living: Criteria for New Construction and Building in Existing Stock, Wiesbaden.

Gardossi, L., et al. (2023a). Bioeconomy national strategies in the G20 and OECD countries: Sharing experiences and comparing existing policies. *EFB Bioeconomy Journal, 3*, 100053.

Gardossi, L., Philp, J., Fava, F., Winickoff, D., D'Aprile, L., Dell'Anno, B., Lenzi, A., et al. (2023). Bioeconomy national strategies in the G20 and OECD countries: Sharing experiences and comparing existing policies. *EFB Bioeconomy Journal, 3*, 100053.

Grefe, C. (2016). Global Gardening. *Bioökonomie–Neuer Raubbau oder Wirtschaftsform der Zukunft,* Bioeconomy–New Exploitation or Economic Form of the Future, Munich.

Grossarth, J. (2024). Was sollten wir unter zirkulärer Bioökonomie im Bauwesen verstehen und warum? Eine Annäherung durch 15 Interviews. *Bioökonomie und Zirkulärwirtschaft im Bauwesen: Eine Einführung* (pp. 337–365). Springer Fachmedien Wiesbaden.

Heisel, F., & Hebel, D. (Eds.). (2021). Urban mining *und kreislaufgerechtes Bauen* and circular building, Stuttgart.

Hodgson, A., et al. (2022a). The US bioeconomy: Charting a course for a resilient and competitive future. *Industrial Biotechnology, 18*(3), 115–136.

Hodgson, A., Maxon, M. E., & Alper, J. (2022b). The US bioeconomy: Charting a course for a resilient and competitive future. *Industrial Biotechnology, 18*(3), 115–136.

Holden, N. M., et al. (2023a). Biocircularity: A framework to define sustainable, circular bioeconomy. *Circular Economy and Sustainability, 3*(1), 77–91.

Holden, N. M., Neill, A. M., Stout, J. C., O'Brien, D., & Morris, M. A. (2023b). Biocircularity: A framework to define sustainable, circular bioeconomy. *Circular Economy and Sustainability, 3*(1), 77–91.

Kircher, M. (2020). *Weg vom Öl: Potenzial und Grenzen der Bioökonomie*. Away from oil: Potential and Limits of Bioeconomy, Wiesbaden.

Lakoff, G., & Johnson, M. (2008). Metaphors we live by, Chicago.

Lanzerath, D., et al. (2022). *Bioeconomy and sustainability*, Wiesbaden.

Lewandowski, I. (2018). *Bioeconomy: Shaping the transition to a sustainable, biobased economy*, Heidelberg/Berlin.

Lewandowski, I. (2018b). *Bioeconomy: Shaping the transition to a sustainable, biobased economy*. Springer.

McDonough, W., & Braungart, M. (2002). Cradle to Cradle, New York.

Müller, A. (2018). *Baustoffrecycling: Entstehung-Aufbereitung-Verwertung*. Building Material Recycling: Origin-Processing-Utilization. Wiesbaden.

Müller, E., & Schmieder, F. (2016). *Begriffsgeschichte und historische Semantik*. Conceptual History and Historical Semantics, Berlin.

Nagler, F. (Ed.). (2021). *Building simply: A guideline*. Basel.

Oxford Analytica. (2024). Farmer protests will disrupt EU trade relations. *Emerald Expert Briefings*.

Pietzsch, J. (2017). *Bioökonomie für Einsteiger*. Bioeconomy for Beginners, Wiesbaden.

Scafetta, N. (2024). Impacts and risks of "realistic" global warming projections for the 21st century. *Geoscience Frontiers, 15*(2), 101774.

Schurr, U. (2020). Food security and healthy nutrition in the context of the bioeconomy. In Pietzsch, J. (Ed.) (2020). *Bioeconomy for beginners, beginners* (pp. 67–75).

Schwarzwälder, H. (2023). *Die digitale Bauwirtschaft—Wege aus der Branchenlogik: Status Quo und Chancen der digitalen Transformation*. The Digital Construction Industry—Ways out of Industry Logic: Status Quo and Opportunities of Digital Transformation, Wiesbaden.

Henkel, A. et al. (2023) Dilemmas of Sustainability: On Relevance and Critical Reflection in Sustainability Research, A Guide, Bielefeld.

Thrän, D., & Moesenfechtel, U. (Eds.). (2020). *Das System Bioökonomie*. The Bioeconomy System, Heidelberg.

Vivien, F.-D., et al. (2019). The hijacking of the bioeconomy. *Ecological Economics, 159*, 189–197.

Wojtynia, N., et al. (2023). Spheres of transformation: Exploring personal, political and practical drivers of farmer agency and behaviour change in the Netherlands. *Environmental Innovation and Societal Transitions, 49*, 100776.

Bioeconomic Construction: Technical, Economic, and Cultural Challenges

2

Abstract

Even though global climate policy is undergoing a significant shift with Donald Trump's election as U.S. President in 2025 and the failure of the "Green Deal" at the EU level consequently, there are several reasons for resource-efficient construction. Resource conservation is a pressing priority in the construction industry, not just for environmental reasons but because the stakes are complex and far-reaching. This article dives into the multi-layered challenge of transforming the sector. First, it explores the global construction industry's profound impacts on the environment and climate. Then, it makes a bold claim: in a pluralistic, democratic society, sustainability cannot simply be mandated. This resistance isn't just about economic pressures; it stems from deep social structures, power dynamics within organizations, and even a lack of urgency among the public. The science is clear: we need rapid, sweeping change. Yet real-world adoption lags behind these appeals, bogged down by systemic inertia. That's why this article focuses on a crucial piece of the puzzle: communication innovation. To make meaningful strides, we need a cultural shift in how we think about building and living. This shift goes beyond green building codes; it taps into concepts like the atmosphere, character, historic value, and aesthetic impact of buildings. How do we bring these ideas into planning? Moderated stakeholder dialogues and participatory approaches offer promising avenues. This overview aims to deepen understanding in a tech-driven field, valuing both technical breakthroughs and the need for realistic transformation timelines. Professionals must realize that people need more than rational facts; they need a genuine cultural motivation to drive sustainable action.

On an autumn day in 2022, a user[1] of the news service later named X (at that time *Twitter*) invited his followers to sketch the way to a climate-neutral construction

J. Grossarth, *Bioeconomy of Buildings*, https://doi.org/10.1007/978-3-031-84014-2_2

industry in three words, many participated. The answers were—besides scientifically negligible polemics—diverse: *Puutuoteteollisuus*—that was the account of the "Association of Wood Processing Industry of Finland"—replied with: "Wood—Circular—Small". In this Finnish case, the preference for wood as a building material was certainly less surprising than that for small buildings—as they also require less wood for their construction. "#Baukultur" was the answer of the account of the German Architects Association "Bauhaus der Erde". The *Degrowth* activist "@LotteMuh" opted for a radical prohibition approach: "New building moratorium. Renovation. Vertical building extension with wood." Others bridged the gap from the question of what materials were being built with to the question of how the construction process was being made possible energetically—fossil or by "green" electricity. After all, doesn't every building material become more climate-friendly if raw material extraction, production and transport are made possible by electricity from so-called renewable energies? "Electrification. Renewable. GEBs," the architect Scott Farbman from Chicago consistently suggested—where GEBs stands for *Grid-Interactive Efficient Buildings*, eco-efficient and smartly controlled buildings that become energy storage. Then a number of specific building materials were also mentioned: straw, hemp, grass, earth, reed, clay. This conversation highlights the complexity that this book will continually address: many concepts, such as degrowth, high-tech, low-tech, and sustainability, among others. But it also gives an impression that resource conservation in construction cannot be discussed differently than in this breadth: technology, economy, energy, materials and building culture.

2.1 Climate Politics as a Turning Point in Construction?

2.1.1 Unsustainability in Construction

Current construction practice is considered unsustainable in the light of science for many reasons (Hossain et al., 2020; Lima et al., 2021; Sev, 2009). In addition to scarcity scenarios of single raw material reserves (Chap. 6 and others), the Chinese or European climate protection goals require drastic changes in the construction industry (Lützkendorf & Balouktsi, 2022). The heating of the building stock in Germany and Europe is to be achieved without net greenhouse gas emissions by the middle of the century, as is the construction process from raw material extraction to erection and demolition. Frequent demolition and rapid new construction are just as questionable as the popular form of living in a detached single-family house (Sánchez-Garrido et al., 2022).

Resource Consumption

Resource consumption in the construction sector[2] continues to rise sharply worldwide. A comparison illustrates Asia's share: China alone used more concrete from 2011 to 2013 than the United States did in the entire twentieth century (Gates, 2021). The world population is expected to grow to 9.5 to more than 10 billion people by the middle of the century (UN, 2022). By 2060, illustrative projections of resource consumption suggest that it will be necessary to build about 500 times the size of New York City to provide urban living space for these people (Glock et al., 2022). This number also illustrates the global dimension of the construction emissions problem, which will hardly be solvable with regional or national sufficiency strategies—i.e., renunciation and moderation regarding apartment sizes or material thicknesses—alone. The globally sharply increasing demand for sustainable living space, in connection with the climate protection goals in Europe and globally, makes the task of climate justice and sustainability in construction seem like the squaring the circle.

Timber Construction and Bioeconomic Innovations

Biomass matters. Total carbon stocks in global biomass are gravely as the total global urban mine of concrete (Fig. 2.1). But there are significant development paths that can make success possible: Increased timber construction appears to be a path to climate-just construction (Wagner et al., 2022) and advances in material use efficiency give hope (Hertwich et al., 2019). Agricultural residues such as straw from cereals or hemp and other lignocellulosic biomasses can also contribute, such as miscanthus (Lewandowski et al., 2000; Awty-Carroll et al., 2023). The use of waste streams such as old wood for building material production (Ferdous et al., 2021) or digital construction planning based on Life cycle analyses (Hollberg, 2020) can be other important factors. After all, the consistent circular management of building materials also contributes to sustainable development, but this implies a changed legal and economic framework, such as binding take-back obligations or mandatory building material passports and demolition instructions (Munaro, 2020). Increases in renewable energy for building operation and especially heating are also relevant for the energy and climate balance.

Resource preservation is a central concern of sustainability research. Technical innovations play a central role in the bioeconomy. Chemical or thermal decomposition processes that expand the usage possibilities for old or residual wood are an example (Sect. 4.7, Chap. 5, Sect. 6.2).

With the growing importance of carbon-binding, renewable building materials, the scientific and political discourse also sees the "thinking outside the box" for the actors in the construction industry virtually as a duty. A bioeconomy of construction therefore also means recognizing the area as a scarce good and raw material competitions as an ethical dilemma (Pinsdorf, 2022; Vogt, 2018). The use of wood as a building material for perhaps, globally speaking, hundreds of millions of new housing units is ambivalent from a sustainability perspective (Mishra et al., 2022). The path of the bioeconomy almost inevitably leads into goal conflicts. An example: Increased afforestation and the use of wood as a building material would be a carbon sink lasting for decades. On the other hand, the removal of wood and more intensive forestry counteracts the carbon storage of forest soils. Other forest functions such as the preservation of biodiversity also need to be considered.

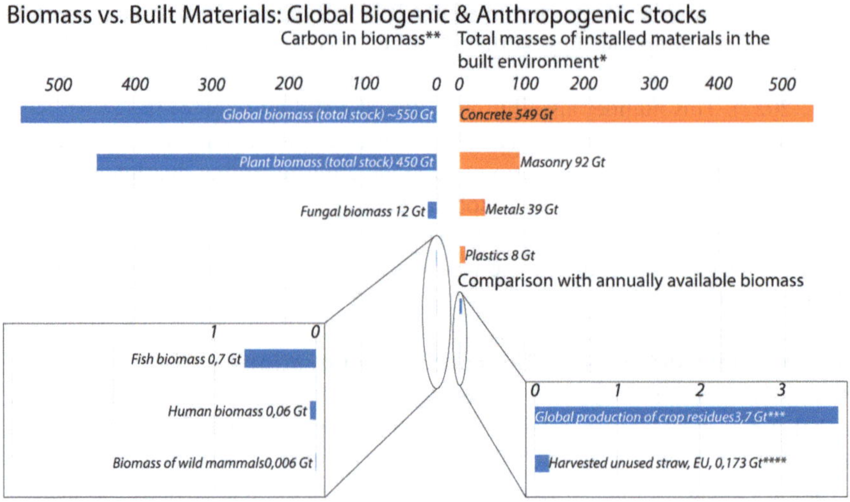

* In 2020. Elhacham et al. (2020). Global human-made mass exceeds all living biomass. Nature, 588(7838), 442-444.** Bar-On et al. (2018).„The biomass distribution on Earth." Proceedings of the National Academy of Sciences 115.25: 6506-6511. *** Before 2005. Li et al. (2023). Optimizing Management Practices under Straw Regimes for Global Sustainable Agricultural Production. Agronomy. 13(3):710. ****2017, of total harvest 0,312,Gt. Göswein, V. et al. (2021), Land availability in Europe for a radical shift toward bio-based construction. Sustainable Cities and Society, 70, 102929.

Fig. 2.1 Biomass matters. Total carbon stocks in global biomass are gravely as the total global urban mine of concrete

Making decisions about the extent of biomass use is a political matter. The sciences of the bioeconomy address goal conflicts, ambivalences, and dilemmas—their concern is not to gloss over these with climate policy or environmental policy platitudes. Science looks at the century project of economic transformation in a value-neutral, descriptive, reflective, and historically informed way.

Global Climate Changes

The 2015 Paris Climate Agreement aimed to limit global warming to well below 2° by 2100 (Dröge, 2015). For this, net-zero emissions of greenhouse gases are considered necessary from the middle of the twenty-first century. Germany has committed to achieving this goal by 2045. The Federal Constitutional Court has defined climate protection as a freedom right of future generations and demands from the legislator that the sector reduction goals be achieved with accompanying legislation. The judgment ties freedom rights to the concept of planetary boundaries (Calliess, 2021; see also Sect. 3.1).

The Paris goal will probably not be achieved. The CO_2 concentration of the Earth's atmosphere has increased from 285 to 421 ppm between 1850 and 2023 (CO_2 Earth, 2023). A further increase in global CO_2 emissions by about 50% by the middle of the century is expected, which would likely put temperatures above the 2-degree goal; the pessimistic forecast scenario by 2100 expects an increase in the global average temperature by 4 °C (Nguyen, 2019). The destruction of terrestrial and marine ecosystems, the thawing of permafrost resulting in further

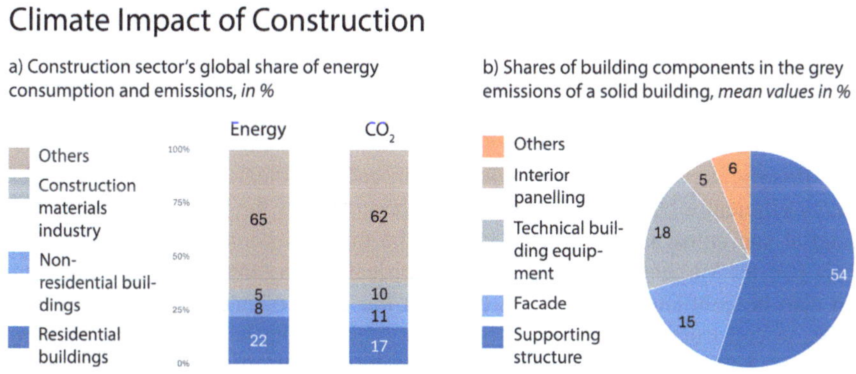

Fig. 2.2 Global climate balances of the construction industry. (JG; Goubran, 2019)

CO_2 release from the soils, the rising sea level, and the accelerated extinction of species are the expected and already observed consequences (Mora et al., 2018).

The construction industry, with a contribution of around 39%, is the largest global emitter of CO_2 (Chen et al., 2023). It is therefore referred to as the "elephant in the climate room" (Schellnhuber, 2021, 2024). The construction industry not only has a major influence on the consumption of mineral and metallic resources, wood resources, fossil and renewable energies, but also on water and land consumption. The construction industry accounts for about 30% of global raw material and around 35% of energy consumption (Goubran, 2019). The largest sources of emissions attributable to the construction industry are residential buildings and, at the building component level, structures, facades, and building systems (Fig. 2.2).

Construction Climate Policy

Not only states determine the speed of the "construction turnaround" through their climate legislation. It is also a concern for municipalities. By the beginning of 2023, 29 cities and metropolitan regions worldwide had committed to climate neutrality in the construction industry, including Heidelberg, Copenhagen, Cape Town, London, New York and Seattle (WGBC, 2023). In municipal networks, the initiative of the construction industry itself is often decisive, as in Malmö.

The political planning perspective expands from the months or few years of the construction process to the entire life cycle of the building and beyond (Gursel et al., 2009). Life cycle assesments in the building sector can also be viewed differentiated by type of use. Worldwide, residential and non-residential buildings contribute about equally to the greenhouse gas emissions from construction and operation over the life cycle—but the significantly higher savings potential can be seen in the area of office buildings (Röck et al., 2020). Digital construction process models (BIM) support this with information that allows relatively easy comparisons of alternatives. The professional world of civil engineering follows the new paradigm in academia and in prestige projects in practice, but the implementation in the broad sense is not yet realised in 2023. Methods such as life cycle assessment are rather slowly integrated into academic engineering education (BauFak, 2023).

The focus on carbon emissions leads to an orientation of the construction industry towards the consideration of the entire building life cycle. In new construction, the emissions of CO_2 (equivalents) for material production and in the construction, process are given more weight than in earlier times. The newer and more energy-efficient a building is in operation, the greater is the proportion of the raw material and construction industry side in the total emissions in the building life cycle (Cortes Vargas et al., 2021). This empirical finding makes it clear that there can also be conflicts of objectives between the goals of energy-saving and low-emission construction. Because if one only looks at the balance of the building operation, buildings can appear environmentally friendly that, on the other hand, either have stored large amounts of "embodied energy" stored in the building or insulation technology, or due to functional or aesthetic aspects are not expected to have as long a lifespan as residential buildings designed according to qualitative aspects.

Interim Summary

Resource-saving construction is necessary. There are many ways to achieve this. The technocratic idea of timber construction as a large global carbon sink raises questions about resource availability and the socio-ecological consequences of global afforestation of plantations. The growth of forests itself is increasingly uncertain to predict due to climate changes (Sect. 4.5.3). There are further risks of climate change related to forest growth, not only the frequently mentioned forest fires (Halofsky et al., 2020). At around 2 °C of global warming, there is a significant increase in the risk of global water scarcity, which would lead to food shortages and intensify the competition for land between building materials and food use (Shukla et al., 2019; Sect. 4.9). Rising temperatures would, however, *ceteris paribus* also have a positive impact on tree growth (Babst et al., 2019). Overall, satellite images suggest reforestation potentials of several hundred million hectares of degraded land, with no major use conflicts to be feared (Bastin et al., 2019). The question of wood resources is hardly clear-cut (Sect. 4.5).

2.1.2 Resource Conservation in Transdisciplinary Research

The focus of climate policy closely links forestry, food, energy and construction material issues. This raises questions like these: Wouldn't it be politically advisable to significantly increase the cost of land-intensive consumption of animal proteins, while making "land-saving" soy or pea protein cheaper? This would free up land for use as forests. So how should renewable building materials be prioritized in the face of land competition? These questions are political, but they illustrate the complex trade-offs that the topic of renewable building materials leads to.

There are agronomic innovations that are suitable for mitigating land competition. Ecological system innovations, such as agroforestry, regenerative agriculture—and simply: agricultural diversification—can contribute to adaptation to climate change and indirectly benefit the construction industry in terms of raw materials (Thevs et al., 2022).

"Economy Without Resource Waste"

A circular bioeconomy means—in the long tradition of this field (Georgescu-Roegen, 1975)—a consistent focus on the energetic foundations of life and economic activity (definitions in later chapters). Many fields of expertise contribute to resource conservation—from engineering, energy technology, agriculture and forestry, plant breeding, biotechnology, but also government legislation. Environmental and waste law belong to the influential circle of urban and settlement planning, but also to the large area that bureaucratizes and legally defines the construction industry: The construction product regulations, fire protection and health protection, countless standards for products, processes and calculations, from statics to life cycle assessment.

To speak of a circular *bioeconomy of the construction industry* does not imply that the "renewable" or "natural"—whatever the latter may exactly be—would have to become the basis of the (construction) economy. Instead, it is about the *guiding idea of closed material cycles*, for which nature provides some examples. Extensive agriculture and forestry merely mark the lower limit of resource availability. The regenerative, renewable biomass is the minimum, the narrowest conceivable energy limit, or in terms of energy history: the "photosynthetic limit" (Niklas & Kerchner, 1984). Resource conservation in the entire economy, in the long term, is the central concern of the circular bioeconomy. The focus is on the technosphere, the mineral building materials.

Longevity as Sustainability Value

The understanding of the longevity of buildings is a good example of the question of what exactly one should imagine under resource-saving construction. What are the key determinants? The factor of the longevity of buildings is gaining increasing weight in the science of sustainable construction, by putting the "embodied energy"—meaning the energy used for raw material extraction, transport, building material production, construction process, maintenance, demolition—of the installed materials into focus (Dixit et al., 2010; Marzouk & Elshaboury, 2022; Skillington et al., 2022). However, greater longevity often comes with technical disadvantages—such as the complex replacement of technical components in the life cycle (Bullen & Love, 2011).

The fact that international climate policy is likely to significantly curb the consumption of gray energy for the production of building materials and construction processes, changes the political coordinates also in adjacent fields. Thus, monument protection sees new justification contexts for an expansion of the mandatory preservation of historical building substance (Guidetti & Ferrara, 2023).

Research Beyond the Disciplines

An interest in the *actual success* of the goal of bioeconomy, a sustainable use of resources, requires *transdisciplinary and problem-oriented research.* The goal of so-called transformative science cannot be to conceal real problems of realisation, and to make vociferous promises to politics. Instead, it must be objective, reality-based, and empirical, with an eye to the meaningfulness of the goal.

Thus, life cycle assessment, which is currently establishing itself as the gold standard of calculation methods for ecological sustainability, is "only" one approach among many. Its circles of experts themselves must face the question of whether the foreseeable dominance of the life cycle assessment approach in the planning process will not result in a technocratic approach to sustainability dominating. LCA usually calculates with standardized lifetimes of buildings, for example according to building durability classes according to Eurocode. However, the goal of increasing the longevity of buildings should—among others—be a concern of the circular bioeconomy of buildings. Creativity in this regard, for example in architecture, is not honored at all in the LCA process. And shouldn't a questioning of the historically high and rising "demand" for living space per capita be a concern of a policy seriously interested in sustainability? Last but not least, human creativity and the joy of design are "resources" that can be inhibited by bureaucratic regulations or technocratically conceived sustainability approaches—as reasonable as all these certainly are to effectively promote resource conservation.

Transdisciplinarity in bioeconomic research

Transdisciplinarity is more than just a buzzword. The concept, which goes back to the work of social psychologist Jean Piaget from the early 1970s, The term transdisciplinarity means that disciplinary boundaries are dissolved in the course of problem-oriented research. This can affect methods or the composition of research or author teams, in which scientists and practical experts, or even "laypeople" work together (Vilsmaier & Lang, 2014). Lang et al., (2012, p. 25ff.) define transdisciplinarity somewhat vaguely as a "reflective, integrative, method-driven scientific principle that aims at contributing solutions […] to societal problems and at the same time to related scientific problems by differentiating and integrating knowledge from various scientific and societal knowledge bases." According to this definition, transdisciplinary research focuses on a socially relevant problem—for example: sustainability in construction. It presupposes a willingness to learn on the part of the researchers, integrating the expertise of non-academics into the research. The latter in particular distinguishes it from interdisciplinarity. Those who opt for such broad transdisciplinary approaches must be able to answer the question: Why would one assume that "lay people" have a deeper understanding of sustainability than academic sustainability experts? Transdisciplinary approaches are popular in the case of sustainability research. The claim of "learning research" refers less to specific methods or procedures than to an attitude. For example, this book incorporates the expertise of scientific and practicing construction professionals through interviews, but also from dozens of my students at Biberach University of Applied Sciences, who have repeatedly responded to my lecture content with questions, comments or (from their practical experience) narratives, leading me to new questions, readings and insights.

The problem of "unsustainability" in general can be seen as a "wicked problem" on many levels. With this term, science describes "evil" or "mischievous" complexly networked, difficult to solve problems (Rittel & Webber, 1974; Vogelpohl & Töller, 2021; Urmetzer et al., 2020). Both for an understanding of the problem and the entanglement of the barriers that stand in the way of a solution, a broad disciplinary view is necessary. The approach of this book sees itself as "problem-oriented research" (Bechmann & Frederichs, 2005), which takes political practice seriously and considers acceptance issues, but also names the relevant natural and environmental scientific problem areas and relates them to the specific concerns of the construction industry. Bioeconomic research presents itself as a model field in which transdisciplinary research must not only remain a buzzword but is downright necessary (Krickhahn, 2022; Scheer & Konrad, 2020). Methodologically, mixed forms of qualitative and quantitative approaches are suitable (Kelle, 2022).

Example

What determines the longevity of a building, and what are the connections to climate-related life cycle assessment? Exemplary questions for transdisciplinary research of circular bioeconomy of buildings

- How does the greenhouse gas-related life cycle assessment of the building change with a lifespan extended by x years?
- How could the factor lifespan be incorporated more flexibly and realistically into the framework of life cycle assessment standards for specific construction projects, instead of being based on average values of building types?
- What criteria determine the actual lifespan for a building of this type?
- How can building lifespans be improved in a cost-effective manner (for example through extensions in wooden module construction)?
- How to predict technological development scenarios for construction in existing buildings or urban mining?
- How can the attention of builders be directed towards aspects of longevity, so that they also influence the decision process?
- Is there a lack of exchange and communication among the actors involved in construction about the aspect of longevity?

- How could such an exchange be *effectively* moderated?
- What are the reasons for demolition? Why do families not choose to extend or renovate the existing building? How does this compare with commercial entrepreneurs? What information could lead to a change of mind? What role do sustainability values play in the decision-making process that determines the demolition?
- Do the relevant actors lack knowledge about options?
- What implicit norms and values are associated with the imperatives of "sustainability" and "climate protection" for the actor groups? And which ones are associated, with the terms "residing", "new building", "homestead", "single-family home"?
- Are there reactances in certain actor circles or social milieus that cause groups of people to close themselves off to the sustainability perspective? What are the reasons?
- What role do negative historical experiences play, but also, for example, conceptual errors? (An example would be the avoidance of wood construction due to fear of fire hazards, which justifies notions of short-livedness-even though technical measurements for new timber materials such as cross-laminated timber bring some salvation in this respect.)
- What role does the beauty of a building play? How do different groups perceive the beauty of existing houses? Or rather: "specialness", "historicity"? And are such properties at all a reason for preservation efforts? At what cost?
- How do empirical-statistical perceptions of beauty differ from each other? Can distinctions be made here between age groups, gender, urban/rural or milieus?

This catalogue of questions is intended to open up the range of relevant questions and their complex interconnections widely, and it could be extended even further. This introductory book only touches cursorily on suitable scientific methods for investigating such multi-layered questions. Instead, it aims to provide an impulse for practice and practice-oriented training, so that they take the concern of problem-oriented research seriously. For research questions, that concern socio-cultural aspects such as "acceptance", "ideas" and "conceptions", the relevant handbooks of empirical and qualitative social and cultural sciences (Baur & Blasius, 2014; Burzan, 2005; Glasze & Mattissek, 2021) are recommended. Whether such research and teaching, which takes complexity seriously, then also contributes to the solution of the problems is another question. At least it can promote reflected practice.

2.2 Cultural and Communicative Challenges in Resource Conservation

The reference to the complexity of the concern and focuses on cultural and social scientific explanational concepts are explicitly not coming from the academic "ivory tower". The broad socio-cultural framework needs to be considered precisely because of its practical relevance. Precisely because of their practical relevance, this essay also proposes terminologies that seem suitable for understanding qualitative aspects of long-lasting construction, such as: "beauty", "atmosphere", "character" (of a building). This expansion is certainly unusual within the strongly eco-balance, engineering or climate science dominated literature on sustainable construction. However, it is also suggested in reference to current literature (Gould et al., 2023; Gkountani & Tsoulfas, 2023) to dare this expansion under the term of bioeconomy of buildings. Then not only the bioeconomy would be an object of study of cultural scientific work—such as its conceptions of "biomass" or "material flow management". But it would be the case that also the social and cultural scientific methods would be part of the transdisciplinary toolbox of bioeconomy itself—especially in actor-centred research (Martin et al., 2023). Narratives and interpretation patterns of the (construction) actor groups are a central research concern (Schmidlehner, 2023). In the broadest sense, it is about an understanding of socially compatible "technology futures" (Grunwald, 2014).

The more the so-called social and cultural scientific "accompanying research" investigates acceptance issues (Hempel et al., 2019; Grossauer & Stoeglehner, 2023), the more one could even ask whether the bioeconomy itself should not also deal with questions of human conceptions of the "good life" (Fehr, 2021)—at least in terms of sustainability (Pyka, 2017; Stein et al., 2018; Winkler et al., 2019). The expanded "toolbox" also promises an understanding of why the concerns of a circular bioeconomy are so difficult to communicate politically and socially (Giurca, 2022).

In the practice of professional sustainability management, the query of technology acceptance through corporate social responsibility (CSR) approaches and stakeholder participation has been well established since about the 2010s (Kussin & Berstermann, 2022; see also later chapters). Most recently, acceptance issues were even proposed in the context of the methodological advancement of life cycle assessment (Rebolledo-Leiva et al., 2023). Where would the points of contact for the construction industry be?

The broadening of the view is necessary if one recognizes the sustainable transformation not only as a technical, but also as a communicative task. The international science of technological impact assessment (TA), for example, is currently undergoing this perspective expansion. It is traditionally engineering-oriented. Authors of TA research now recommend the hermeneutic analysis of societal *interpretations of technologies*, in addition to the established impact assessments (Grunwald, 2020). Thus, not only *factually* expected consequences of technology, risks, dangers and opportunities become research concerns—but also the *thinking and speaking* about it (Wittmayer et al., 2019; Schöggl et al., 2020). Texts or works of art can be objects of technology-related hermeneutics—but why not also architectural designs of sustainable quarters? One could show the new model house designs in draft to various citizens and landscape architects and ecologists and other "stakeholders" (again—a term established in science and practice, but terribly technicist, because who would actually want to call themselves that?) and use their feedback for improvements.

Considering cultural conceptions and social values can be of practical interest for technically and economically oriented actors in several ways. It allows, first, conclusions about the acceptance of a "transformation" towards climate-just construction, second, the communicability and marketability of corresponding buildings or real estate. Third, it allows the technical world or the construction industry to look in the mirror.

Because many questions are in the room when talking about "transformation" and its acceptance: Is the politically majority-capable path of sustainability and climate neutrality of the economy really realizable (Siegel et al., 2020)? To what extent is this claim compatible with the anthropological reality of humans, who also want to "grow", shape and build (Goh et al., 2020; Pelicice & Castello, 2021)? Does it fit the reality of a world where scarcity and war prevail? To one where the model of autocracy has regained attractiveness? Can this claim be met in a political reality where nationalism is on the rise again and the liberal world order is threatened to fail (Colgan & Keohane, 2017; Fukuyama, 2022)? How can the political framework be effectively and sustainably aligned (Tomor et al., 2019)?

The science-driven "planetary" orientation of politics seems necessary given the complexity of global crises, but politics must be measured against regional acceptance and majority will. In this context, questions of concepts of freedom and restrictions on freedom also ariseAn "ethical discussion" would address related dilemmas: How do freedom and protection rights relate to each other, consumer freedoms to entrepreneurial freedoms? (Schleissing, 2022) One can certainly view sustainability merely "technocratically"—i.e., within the narrow framework of legal goals that are to be solved technically and politically. But one can also see the path towards sustainability as a communicative process. In this framework, even if the goals prove to be (still) unattainable, an orientation towards sustainability would make sense, as it, for example, orients the creativity and innovativeness of human freedom in this direction. This is the significance of a wise choice of words. Precisely for this reason, the bioeconomic view, as it is to be stimulated in this book, means that not only the large statistical target quotas, but every successful conversation, every pleasing building that promises greater longevity and inspires imitation, makes a relevant contribution.

Through houses, people interact with their environment. Building is not just a technical process that consumes resources but is itself communication. Every characterful, protectable and therefore long-lasting house is a testament to effective, unintended, practical resource conservation. Building as a "communicative action" (Knoblauch, 2016) influences the health of the environment and humans. It shapes human living spaces—just as, for example, agriculture shapes the landscape. The result is either talk of cultural landscapes or industrial agriculture, of orchards or monocultures. Similarly, in the case of building, we speak of apartment blocks, quarters, residential silos, hotel complexes or single-family homes. These are cultural interpretations.

This is not about "conscious communication"—or even strategically planned—through building designs and implementation. Even the functional, affordable, legally compliant and energy-saving worthy construction is a "communicative act". With this, a builder communicates these characteristics. Even what was not built is communication (Watzlawick, 2015). Sustainability is on the one hand an assertion. It is associated—at least in the media—with certain building materials or styles (wooden construction), claimed and denied, calculated and certified. On the other hand, sustainability is an interpretation, no matter how elaborately quantitatively substantiated mathematically, no matter how eloquently justified. The life cycle assessment or certificates for sustainable construction suggest an absolute objectifiability of sustainability. But they are based on methodological and calculatory determinations, all of which did not "fall from the sky"—and leave a lot of room for design. But they obscure this to the extent that they simplify complexity.

2.3 Aesthetics and Sustainability

> A beautiful house is loved by people, and such a house is more sustainable than any other in the world. Cameron Sinclair (Archdaily, 2020)

2.3.1 Appreciation of Beauty and Sustainability

An example of the relevance of qualitative research on this is the value of the perceived beauty of a building (Triantafyllidis, 2001). Feelings of beauty can be surveyed. In the case of the perception of beauty of different landscape types, for example, it has been empirically shown that open, savannah-like landscapes are predominantly perceived as particularly appealing (Pierskalla et al., 2016; Tveit et al., 2018). In the case of buildings, symmetrical lines (Schummer, 2006), harmonious and calm shapes, whose use in a broad historical retrospect is subject to changing fashions and developmental dynamics of interpretation (Bisky, 2000; cf. Fig. 2.3).

Consideration of aesthetic impact is occasionally even described as a core concern of ecological "transformation" (Hosey, 2012, p. 8ff.). Judgments about it are by no means subject to subjective arbitrariness (McMahon, 2005) and in the case of buildings are scientifically categorized, for example, based on the reactions of viewers of images (Lavdas & Salingaros, 2022). Even if this was not the original intention of this discipline, the work of landscape architecture oriented towards beauty also contributes significantly to ecological sustainability (Meyer, 2008).

In sustainability certifications such as BREEAM, aesthetic factors are also considered, but heavily underweighted and only indirectly addressed (in categories like

Fig. 2.3 Beauty through simple means: A facade in southern Denmark, which has a calming and balancing effect. Village images: The historic town of Buchen in Germany's Odenwald with half-timbered houses—and a village in the Lombardy, where the village and surroundings are clearly demarcated, instead of the often sprawling urban sprawl due to new construction. Photos Jan Grossarth, Postcard Delcampe

"materials", "design and performance" or "health and wellbeing). Even the explicitly "people- rather than environment-centred" American label WELL does not focus on aesthetic aspects. This has been criticized: "This leads to the conclusion that sustainable architecture must contribute to social justice, aesthetic qualities of the environment and the preservation of cultural values. Therefore, aesthetics must be considered as an integral part of architectural sustainability." (Grazuleviciute-Vileniske, 2021, p. 85ff.). One should add: Although the value of beauty in art and philosophy has somewhat fallen into disrepute in the decades following World War II, alongside ornament and stucco (Liessmann, 2009, p. 49).

2.3.2 Buildings as Communication

If building processes are viewed as communicative actions, it follows that a contemporary concept of communication should be employed. Communication is not a one-way, linear transmission of information from sender to receiver. Perception and interpretation are personal and experience-based; they are socially and culturally determined, robust, and self-reproducing. Grunwald refers to this as the 'hermeneutic circle' (Grunwald, 2020).A widely established communication scheme—the so-called iceberg model—distinguishes the factual (information) from the relationship level. The latter is largely decisive for understanding. Helpful methods such as Life Cycle Assessment (LCA) would be located on the factual level. This also includes data on durability, energy- and CO_2 balances, as databases like Global *LCA Data Access Network* or *Ecoinvent*, or building specific German *Ökobaudat* or *Wecobis* provide. Because this perspective is helpful, but alone it falls short to approach the goal of sustainable construction adequately problem- and practice-oriented, the iceberg model is helpful (e.g., Motschnig & Nylk, 2009, p. 46ff.). Below the surface in this model are the majority of factors that determine the "success" of communication (Fig. 2.4). The feeling that the sight of a building triggers in people (one, several, the majority in representative surveys) would be located here. Aren't these factors crucial because they decide on the lifetimes of buildings in each individual case? A narrow focus on "beauty" is not at all useful. Originality, uniqueness, association, memory, rarity, coherence, contemporary historical factors would be further relevant factors that could be located "below the surface".

When we talk about "culture", we mean culture as a *pattern of orientation in the social.*

> Culture is built on ideas, patterns, and values, so it is always historically conditioned because it is learned. [...] Culture is the abstract basis of behavior and at the same time the product of behavior.
>
> Gunther Hirschfelder (2012a, 2012b)

Culture is also defined as "fabric of meaning in terms of which human beings interpret their experience" (Geertz, 2000, p. 145). Here, experience is already central by definition.

2.3.3 Architecture and Cultural Semantics

"Building" in Old High German ("buan") was still synonymous with "living", i.e. staying in the building, being *there.* The "cherishing and nurturing", Latin

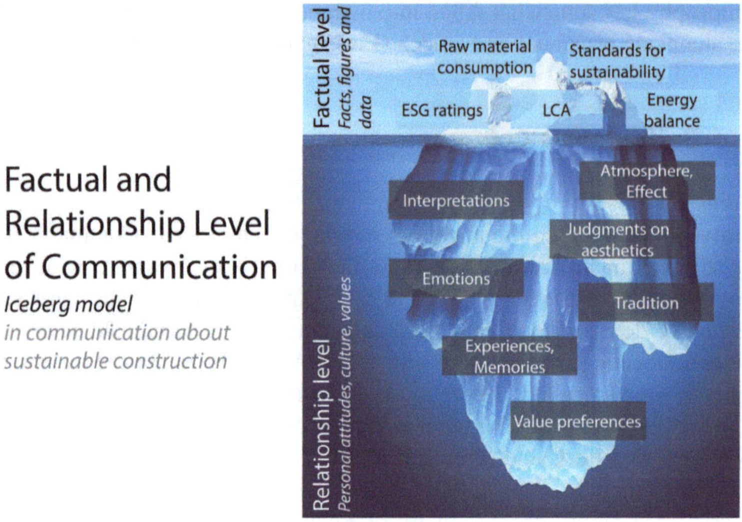

Fig. 2.4 The iceberg model of construction communication. JG

"colere", as it still resonates in the word agrarian culture as "land cultivation", actually also means "building", "inhabiting", "caring", "revering", "sanctifying" (Pons, 2023). The culturally positive semantics, which is still contained in the word "land cultivation", can hardly be traced in the word construction industry—but quite in the word construction culture. Subjects of construction-related cultural research include:

- Urbanity
- Construction culture
- Living culture
- Ecology
- Corporate culture
- Related ways of thinking, mentalities.

2.3.4 Narratives in Sustainability

Objectifications of sustainability—whether in the catalogues of criteria of certification bodies, in construction standards or in academic definitions—are the result of ideas, interests, positions and negotiations about them. Even technical terms have a history (Müller & Schmieder, 2016). Acceptance of technology is not primarily determined by technical features. It arises in a discursive "struggle for interpretive authority" against the background of majority-capable and "powerful" narratives (Traue et al., 2022). Is agrochemistry a poison or a progress? Is nuclear power climate-neutral or deadly? Is concrete outstandingly durable or a sin against sustainability? Debates about this are in their media Conditionality, their scientific versatility, and their historicity to understand and evaluate accordingly.

The art of scientific interpretation is hermeneutics. In this sense, sustainability, bioeconomy, and circular economy are also concerns of the scientific disciplines of comparative cultural studies, historical science, discourse analysis, and literary hermeneutics (Grunwald, 2020). Sustainability can be interpreted as a narrative—as a meaningful story.

2.3.5 Cultural Semiotics in Construction

One can thus look at the documents and speeches of engineering congresses; the interview with architecture students as well as Instagram posts by climate activists, or the brochures and advertisements of housing construction corporations that advertise sustainability. The practical value lies in the fact that this enhances the understanding of the complexity of communication—and thus the "solvability" of complex problems such as climate change or species extinction appears in a more realistic light. It is also not just about moderating expectations, but better success in intercultural communication (Heringer, 2015). In this sense, one can also scientifically discuss aspects of building such as beauty, character, coherence, and effect, without meaning this normatively (Böhme, 2013). Perhaps this also includes the historicity of a building, the interesting thing that results from temporally, locally, and biographically assignable symbolism. Such *cultural semiotic approaches* are already traced back to the Italian philosopher Giambattista Vico. In this sense, beauty would be understood as something like an "inner" beauty, which results from the "readability" of a building.

An understanding of building as a communicative action finds correspondence in cultural anthropological, psychologically founded approaches that refer to Vico:

> People anticipate, invent, and pre-a d apt through culturally suggested imaginative, narrative, and creative processes that lead to the production of meaningful actions and artifacts. These processes are not just intrapsychological or psychophysiological. They are also socioculturally embedded in everyday historically situated contexts of activity, following dialogical, imaginative, and narrative logics. (Tateo, 2017a, p. xiii)

An example of an enlightening work that shows a cultural change in construction is an analysis of housing construction brochures in a long historical perspective. From this, an interpretation of a change in the ideas, concepts, and orientations of the planners, builders, and marketers can be derived. For the Basel area, such a qualitative content analysis found, for example, that from about 1870 to 1945 the social class question was crucially expressed in the brochures or advertisements for new buildings (Kriese & Scholz, 2012). The move from "tenement blocks", from the "noise and dust" of the cities, into terraced houses in "light and air" of the suburbs was described here in the context of the social class question. A fundamental change to the justification logics of a lifestyle society can be identified from the 1950s onwards. The "own home" is now advertised as a symbolic code for family-related values, but also the retreat from the public to the private of the petty bourgeois family (Warda, 2020). While industrial corporations in the pre-war period, led by patriarchal figures, still built generous housing estates for their workers to contribute productively as an actor in the class question, from the 1950s onwards "you make your own luck". The advertisements of the housing projects bear witness to this (Kriese & Scholz, 2012): In the 1980s, possibilities of individualisation of pre-planned terraced houses in the countryside come to the fore of the presentation in the brochures. From the 1990s onwards, it is clichés of happy country life, the "pure living pleasure" or later of the bright floor-to-ceiling window or similar. Studies like this express that now the marketing logics of an individualized mass consumption society prevail. The dominant influence of consumers is discussed as power-critical theory under the term consumerism (Baum, 2022; Chase, 1991).

Atmosphere

The concept of atmosphere used in the philosophy of landscape aesthetics, refers to a relationship between observer and object—referring to landscapes, but also to buildings (Böhme, 2013) "In any case, the reality of atmospheres requires a physically present human being who, in his state of mind, feels in what kind of space he is," writes Gernot Böhme (2020; see Fig. 2.5). In his standard work, he defines the atmosphere as follows: "The atmosphere is the shared reality of the perceiver and the perceived. It is the reality of the perceived as sphere of its presence and the reality of the perceiver, insofar as he, sensing the atmosphere, is physically present in a certain way" (Böhme, 1995, p. 34). Böhme sees in the concept of atmosphere an extension of the concept of aesthetic judgment, which goes back to Immanuel Kant, towards a concept of "sensuous knowledge" (Böhme, 2020).

Fig. 2.5 The philosopher Gernot Böhme (1937–2022) during a conversation in 2021. Photo by Jan Grossarth

Discourse Analysis

Engineers tend to rely on the persuasive power of the "factual". But there are many competing ideas about sustainability. Which ones "prevail" politically or scientifically—for example in standards or certifications—is the result of discourses about it. Discourse analysis is the examination of an argumentation-exchange with a structural-analytical view of the claims to truth and reason conveyed therein, as well as the question of their legitimisation. This approach goes far beyond a mere argument analysis. The French philosopher and sociologist Michel Foucault points out that discourse analysis is about "reintroducing the rhetoric, the speaker, the debate into the field of analysis", in order to "examine the discourse about truth as a set of rhetorical procedures" (cf. Grossarth, 2018, p. 48; Foucault, 2002, p. 779).

Hermeneutics

Hermeneutics is suitable for analyzing the ideas of sustainability of various actors—be it of texts, or of buildings. There are numerous scientific approaches to hermeneutics, ranging from "objective" to "poetic" to legal. Hermeneutics is "sense reconstructing sign interpretation" (Kurt, 2008, p. 369). As an art of interpretation, it provides a multi-layered classification of sources. These can be the time or life history of the authors, aspects of the history of ideas (of architectural history), historical events or a language and image analysis with a view to stylistic devices, symbolism/metaphorics or—in the case of text sources—histories of concepts or etymology of words. A classification in the history of effects is relevant. The goal is the research and "reconstruction of individual and societal sense constructions", whereby *not complete understanding* of the intentions of the authors or the meaning content of the source is the point of the hermeneutic approach, but a "typifying" relationship of the human-individual to the societal and historical level. Two central assumptions of hermeneutics are: Similarities and equality of human thinking and interpreting can be (firstly) understood, although (secondly) "man is a stranger to man" (Grossarth, 2018, p. 52; Kurt & Herbrik, 2014, p. 477ff.). The hermeneutic analysis is also applied to "socio-spatial phenomena" such as buildings. Social science hermeneutics is suitable as an "approach to socio-spatial problems" such as housing; it has, for example as "example hermeneutics", its advantages in local, exploratory theory-forming concretion (Görgl, 2008, p. 42ff.). In Blumenberg, it is also fruitfully said for the "reading" of bioeconomy undertaken here: "Hermeneutics goes to what not only should have and reveal a meaning for all times, but what precisely because of its ambiguity incorporates its interpretations into its meaning. It assumes that its object enriches itself through constantly new interpretation, so that its historical reality lies precisely in accepting new readings, carrying new interpretations." (Blumenberg, 2020 [1986], p. 21).

2.3.6 Human-Technology Discourses

Hermeneutics and discourse analysis allow conclusions about the relationship between humans and object/technology. The technical approach can be as intrusive as the surplus of meaning (Paddock, 2004, p. 239f.). The power or tragedy of the engineer is a major theme in literature, especially in the first half of the twentieth century. The "homo technicus" or "homo oeconomicus" become technocratic figures of misfortune in literature (Blaschke, 2004; Irrgang, 2010). The dominant role of engineers and the almost unlimited political trust in progress until about 1970 is also critically interpreted (Fraunholz & Wölfel, 2012). Around 1970 is the epochal change that began the "age of ecology" (Radkau, 2011).

Ecology is a way of so-called holistic thinking. Its theme, the devastating split between technical and emotional reason (Meier-Seethaler, 2001)—can also be transferred to areas of construction. This dilemma is neither to be solved nor discussed in depth here. However, such a broad view of the problem of resource waste is relevant—also to understand the backgrounds of the media and political topic cycles of sustainability. Because it is not only the scientific crisis findings of climate change and biodiversity losses that have caused this (Uekötter, 2012; Grossarth, 2018). A political motive is the discomfort about what becomes of man in the technical age, or whether one should even speak of "technology dominance". Emotional reactions to buildings can be described as *resonance*, and the scientific term atmosphere or *felt sense* can be helpful for their scientific handling.

2.3.7 Building Culture

Living and building forms can be viewed culturally as communications of understandings of identity. The use of historical building materials could, for example, be understood as a traditional anchor. Aspects of building culture are comfort, durability, and beauty. The concept is not sharply defined and is closely linked to ecology as a relationship science (Furrer, 1992). Building culture is opposed to purely functional approaches. In the case of agricultural farm buildings, for example, a stronger consideration of beauty and the effect of the buildings with a view to societal acceptance problems is recently suggested (Dauermann & Enneking, 2019). In 2018, all EU member state ministers for art declared in the Swiss city of Davos that Baukultur (building culture) would be sustainable. They referred to a form of high culture (Hochkultur) in buildings, but the declaration was vague and scientifically unsatisfactory, and it has been critically discussed since (Grandits & von Planta, 2022; Schnell, 2022).

2.3.8 Emotional Responses to Technology

One of the core observations of the Frankfurt School was that in a mass consumer society, the personal passion for design of experts involved in the industrial process hardly has any place or validity. The corresponding well-known criticism of Theodor W. Adorno, also of the "culture industry", could easily be transferred to the construction industry. He saw, for example, music or writing reduced to industrial goods through publishing marketing and media attention mechanisms (Schweppenhäuser, 2019). A predominance of "instrumental reason" (Max Horkheimer) was lamented. In this tradition stand updated critical thematisations that describe the "reduced existence" of humans in the industrially digitized modernity (cf. Friesacher, 2022, pp. 57–66).

Recently, Hartmut Rosa's concept of "resonance" should be mentioned here, which he presents as the core of a "sociology of world relationship" (Rosa, 2016, p. 20). Rosa refers to the communicative relationship between human and object.

Resonance means something like the "touch by the unavailable other" (Rosa, 2016, p. 621.) Further current writings in the tradition of the "Frankfurt School" deal with ideologies of performance or populism, which is due to the fact that even truth has become a commodity (Distelhorst, 2014, 2019). Ecological crisis descriptions can also be hermeneutically interpreted as ciphers of such cultural-critical perceptions (Grossarth, 2018).

2.3.9 Felt Sense

The psychoanalyst Eugene Gentlin recommends this term for a *felt, perceived meaning*. This is, for example, the emotional resonance to a problem that is revealed in conversation, to a vision of the future, but also when looking at an object, such as a building (Wiltschko, 2000, p. 205f.). Gentlin says: "Such a bodily feeling can be obtained for any thing: for a small or large, for an unpleasant or pleasant, for one from the past, from the present or the future. This 'thing' can be an idea or a situation or another person or a problem. The 'thing' can be anything. You can think anything about such a thing, but you always have a bodily feeling for this whole thing. This bodily feeling is a good guide. I call it Felt Sense." (Gendlin & Wiltschko, 1999, p. 2) This feeling comes first, followed by rationalisation. Other authors speak of "reasoning" (Habermas & Köber, 2015) and "sense making". Related to these differently accented approaches to embodiment is Thomas Fuchs' concept of "*Embodiment*" (Fuchs, 2020, 2021). The complex connection of feeling and thoughts described therein exposes established notions of a separation of brain and body, of reason and feeling.

2.4 Economic and Technical Challenges in Resource-Saving Construction

2.4.1 *Economy:* Division of Labor and Modern Society's Inertia

> For example, there is a process of rethinking [towards sustainability; JG] in the management of the chemical industry, as we know from interviews, and the question is how this will affect the market. Will it necessarily lead to higher debt rates and lower earning opportunities, which will then limit the continuation of this very policy? If you don't have money, of course, you can't do anything. These are the really interesting questions, and I am first and foremost an unbiased observer. And I believe that these are the crucial points.
>
> Niklas Luhmann in response to an interview question (Luhmann et al., 1990, p. 27)

Sustainability incurs business costs, non-sustainability economic costs in the long run. Nothing has changed in this dilemma since 1990. Luhmann's quote illustrates in retrospect, how long "one" has been speaking in a comparable way as in the present at crucial points in economy and politics about the concern and urgency of sustainability. The sociologist Niklas Luhmann, as a theorist of modern society, has provided the conceptual approaches to understand the reasons for this.

Modern society is characterized by increasing individualisation of lifestyles, rationalisation of production and an increasing functional differentiation of social areas such as economy, politics, science and religion (van der Loo & van Reyen, 1992; Braun, 1993; Luhmann & Heidbrink, 2004). An orientation towards sustainability would be overarching. Indeed, the high degree of division of labor, specialisation and differentiation of tasks poses a problem for the "alignment" towards sustainability. Modernisation cannot be simply revised. There are reasons for the development, especially economic advantages of rationalisation and division of labor and corresponding institutionalization—for example, the increasingly specialized training and study programs. The economic history, which has brought about these developments and the economic growth as *interconnected*, is based on linear value chains that lead to waste generation, environmental damage and resource shortages.

In the field of economics, behavior is primarily oriented towards binary codes of monetary values such as price, revenue, return (Luhmann, 1984). Which builder paid three times the price for a rammed earth wall compared to a solid wood wall? Economic factors, however, only become significant for specialized professionals in purchasing or sales who operate within given budgets. They are also of great relevance to executives when they are under pressure to publish quarterly figures, which affects stock prices, their own remuneration and career prospects. Playing hardball in price negotiations is the craft of professional buyers. The bonus and remuneration systems in trading and manufacturing companies and also the construction and real estate markets are aligned accordingly.

The sociological system view provides an understanding of the "inertia of transformation" (Nassehi, 2021). It brings the insight, that functional differentiation essentially maintains and solidifies itself. Actor groups in the sub-areas of society therefore primarily orient their actions according to "binary codes" (according to sociologist Niklas Luhmann). The economy, technology, communication and the judiciary have their own specialized expert groups, own action logics and a limited understanding of the other perspectives—at least this would not be relevant to action.

Ecological communication

The sociological system theory describes the order of modern society. It also explains why outrage and increased appeals to the economy that it must become transformative as soon as possible, have borne little fruit. Even media enhanced moral accusations against managers and against ideologies of the homo oeconomicus lead according to this theory to no resonance but are merely perceived as noise without informational value. It is quite possible that even the addressees emphatically agree—but the next day in the office they act again as actors of the "functional area economy" in a contrary way.

That modern society is functionally differentiated is based on historical and technical development as well as constant differentiation such as progressive division of labor in the economic or separation of powers in the political system. But modernisation is ambivalent. With it are not only disadvantages such as unintended ecological follow-up costs, but also advantages associated with increasing productivity of the economy, in the case of competitive market economies high price pressure, which leads to affordable goods for larger circles (social aspect of sustainability). The ambivalence with ecological demands for wholeness cannot be "healed" to transformation by good will. This analysis and the concept of ecological communication itself go back to Niklas Luhmann, are not new, but in the context of increased political transformation efforts, they are worth remembering again (Luhmann, 1993).

2.4.2 Do Sustainability Campaigns Change Consumer Behavior?

It would be naïve to think that more "education" and information about scientific findings on the climate and environmental consequences of certain construction techniques could significantly change the actions within the given corridors of such economic actors. But do educational campaigns work for consumers by influencing their actual buying behavior? Certainly, they also have a staging function for the senders. But higher value creation can be achieved through sustainability information at the "*point of sale*"—at least in the consumer goods sector—as the success story of high-priced organic labels in the food sector shows. These are not only informative, but also have the character of brands. For the construction industry, there are hardly any relevant studies on how sustainability information affects consumer decisions.

But there are research findings from the comparative field of nutrition. Nutrition-related sustainability campaigns do have an impact on attitude changes towards sustainability, but hardly cause behavior change (Godfrey & Feng, 2017). Social reference groups probably have a greater influence on behavior than information campaigns in traditional media (McKenzie-Mohr, 2011). The predominantly published psychological literature or marketing literature on this topic focuses more on "types" and "values" of consumers, economic factors of consumption are not considered (Zhuang et al., 2021). In addition to price, everyday routines also stand in the way of a consumer-driven turn towards sustainability (Schäfer et al., 2012).

Under the psychological and behavioral science concept of *nudging*, a changed placement of products or targeted information at the point of sale is understood, in this field behavior changes have been demonstrated especially in the area of gastronomy (canteens) or in food retail (Vandenbroele et al., 2020). The suitability of techniques of "soft" market influence in the construction industry—such as real estate marketing—is scientifically hardly examined so far.

The action effect of information is often overestimated. For decades, biochemically and nutritionally oriented expert circles have relied on the power of information. They bring nutrition pyramids and display boards about the right proportions of meat, milk, grain, vegetables, and fruit into the textbooks—with the result that this information is now widely known, but hardly gain relevance in everyday life (Spiller & Nitzko, 2021).

Cultural studies and social psychology have provided explanations for this failure of nutrition education (Barlösius & Rehaag, 2006; Hirschfelder & Pollmer, 2020; Spiekermann; Pudel; Godemann & Bartelmeß, 2021). They show the high explanatory value of the socio-cultural view: Nutrition is not just food intake, but eating itself is a cultural act, which is essentially determined by memories, traditions, affects (pleasure and displeasure, frustration) (Hirschfelder, 2012a, 2012b). For example, participants of obesity self-help groups have the best knowledge about healthy nutrition but explain fatigue and lack of motivation after a tiring workday and long commute as reasons for not eating the salad with fruit in the evening, but the quick greasy snack (Hitzler, 2020). How could such insights be translated into research questions that concern the construction industry? Which orientation values, attitudes, references to social reference groups, could decisions for sustainable building give consumers in decision-making situations?

2.4.3 Actor Groups in the Construction and Building Technology Industry

A schematic sketch of the actor groups and the economic relationships among them in the construction process illustrates the complexity of a "transformation" towards sustainability (Graphic 2). In such a process chain, price negotiations occur at dozens of levels. The sketch also indirectly gives an impression of how important the regulation of the financial market framework and the non-financial disclosure obligations (EU taxonomy) are if politics wants to effectively orient economic processes towards sustainability. Economic action seems to dominate, as it is effective at so many levels. The sketch also suggests that general contractors, architecture firms or the raw materials trade are central hubs regarding the use of resources. And it also becomes clear that the buyers of real estate often are the last link in a chain of decisions made primarily on economic criteria. Normally, they only come into play when it has long been decided that sustainability in the construction process has played little or no role.

2.4.4 Value-Chain Communication

That's why it's so important to initiate, moderate and maintain a change in the way economic actors communicate with each other. It must be started early and accompanied by concrete sustainability goals. Such *Value Chain Innovations* hold the justified hope that the powerlessness against the economic motive does not have to have the "last word" (De Angelis et al., 2018; Gupta, 2020). But even such project-related cooperating value chain actors are—according to Luhmann—probably not a "panacea" for the entire national economy. They are a new and interesting path, for which there are still few lighthouse projects from the construction industry. It requires coordination and moderation (Qu & Liu, 2022). Digital communication tools and *blockchain* can accelerate them (Bhattacharya & Chatterjee, 2022).

The abstractions of systems theory and the emphasis on binary codes—which cannot be subsumed under a higher societal rationality—of decision-making are to be understood in this sketch as model-like simplifications. In addition to functional differentiation, there are many approaches to explaining the inertia of sustainability development. Social role behavior or conformism are among these, the latter is a term from social psychology of actors' interactions(see Fig. 2.6).

2.4.5 Conformism—Conservatism—Power

Unlike the sociological, almost mechanical view of "systems", social psychology focuses on human irrationality, which is associated with the pleasure or affinity for exercising power. It often proves these experimentally (Gerrig & Zimbardo, 2008). Such aspects are important additions that can justify more democratic decision-making, stakeholder involvement in operational or strategic corporate decisions

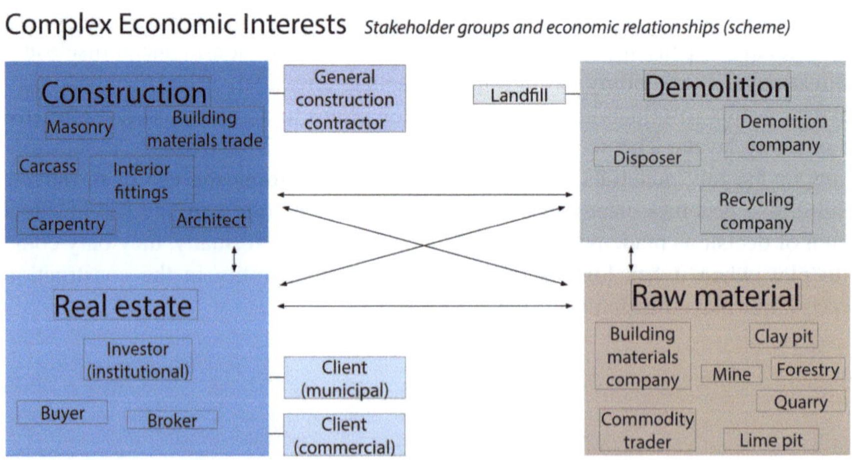

Fig. 2.6 The actor groups involved in the construction process. (JG)

(although the latter also bring with them new, different power structures). Motives such as power expansion, power retention or prestige gains can have a significant influence on economic decisions. Experimental social psychology reveals powerful human irrationalities in the economic process such as

- Conformism
- Conservatism
- Self and other attribution of achievements or failures (the tendency is to blame failure on other people, attributing successes to one's own performance)
- Lust for power, cruelty, desire for self-promotion.

The significance of social psychology for understanding economics has been acknowledged (Schlicht, 1998). A further look at the peculiarities of human perception in economic activity is provided by behavioral economics (Beck, 2014). In addition, institutional economics, as an analytical level of consideration, relativizes the dominant importance of purely economic decision factors. It focuses on formal and informal institutions such as rights, laws, norms and their economic significance (Hodgson, 2004). The terms of all these different scientific approaches can contribute to explaining the problem area of non-sustainability in economic action and the inertia in a transdisciplinary way—but also how "sustainability hypes" emerged in large circles of the economy at the beginning of the 2020s (which do not necessarily have to last long and do not necessarily result in effectiveness).

2.4.6 Technological Innovation as a "Behavior-Neutral" Sustainability Lever

On many levels, from mechanical engineering for the construction industry to framework technologies of the building materials trade, technological development indirectly changes the decision-making basis of economic actors (Fig. 2.7) through altered costs. Higher energy efficiency, resource efficiency or more climate-friendly materials can be the result of technical development. In anticipatory, comparative life cycle assessment studies, technical development has the greatest "sustainability potential" for the construction industry (Karlsson et al., 2021).

Technological innovations are therefore of great importance because technical development can lead to ecologically more sustainable action without the burden lying on practically irrelevant hopes such as "value change", "moral conviction" and "new behaviors". For example, not all people would have to subscribe to a morality of renunciation if meat substitute products made from oats, peas or the petri dish were nutritionally, tastefully and haptically identical, but cheaper than meat. Then one would no longer have to say: Sustainability is important—but it's the economy, stupid. However, technological development does not fall from the sky. Societal expectations of the future form its basis, especially nations set the incentives and the industry ultimately develops the techniques.

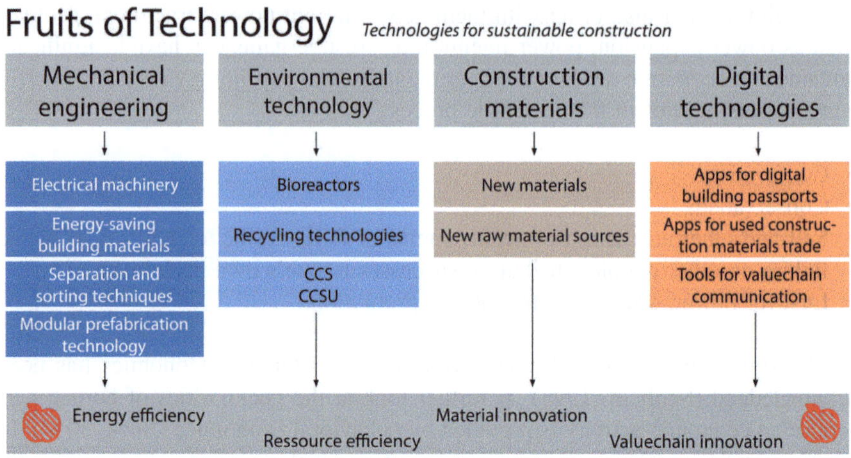

Fig. 2.7 Technical developments influencing sustainability in the construction industry JG

2.5 Summary

Resource conservation is desirable for many reasons.

This chapter has addressed the complexity of the transformation project. It first listed the severe environmental and climate impacts of global construction, then explained why greater sustainability cannot simply be regulated in a pluralistic and modern society. The socially differentiated structure, power aspects in organisations, and the reactance to ecological narratives are three examples of the inertia of transformation despite acute planetary crises. Sustainability was described as an ideal that can be approached primarily through technical progress and communicative innovations. A central role is played by digitally supported, moderated agreements of all relevant actors in the construction process.

Finally, the chapter also contains a plea for a cultural opening of the technically dominated construction sector. How can the beauty and character traits of houses be described and assessments compared to give them validity in the planning process? Here too, a cultural and communication science informed exchange in the form of professionally moderated so-called stakeholder dialogues and other forms of participation is suitable for practice. The overview is intended to provide an adequate understanding of communication processes in a technically trained professional world. On the other hand, it should appropriately define the expectation framework of the transformation. As motivation for sustainable action, people need more than abstract insight into the necessity.

Notes

1. Tweet from Michael Burchert, @BuchertMichael, from October 5, 2022: "Name 3 keywords to make construction climate neutral".
2. A "construction sector" is not recorded in the official German climate statistics, which makes the interpretation of the statistics presupposed. This economic sector is found on the one hand in the industry sector—this includes, for example, building materials—, corresponding reduction goals apply. On the other hand, there are GHG reduction goals for the building sector, where the focus is essentially on heating operation.

References

ArchDaily. (2020). Cameron Sinclair. Retrieved February 13, 2023, from https://www.archdaily.com/tag/cameron-sinclair

Awty-Carroll, D., et al. (2023). Yield performance of 14 novel inter-and intra-species Miscanthus hybrids across Europe. *Global Change Biology. Bioenergy.*

Babst, F., et al. (2019). Twentieth century redistribution in climatic drivers of global tree growth. *Science Advances, 5*, 1.

Bar-On, Y. M., et al. (2018). The biomass distribution on Earth. *Proceedings of the National Academy of Sciences, 115*(25), 6506–6511.

Barlösius, E., & Rehaag, R. (2006). *Skandal oder Kontinuität: Anforderungen an eine öffentliche Ernährungskommunikation* (p. 117).

Bastin, J.-F., et al. (2019). The global tree restoration potential. *Science, 365*(6448), 76–79.

BauFak. (January 9, 2023). *Bauingenieur-Fachschaftenkonferenz, Positionspapier Nachhaltiges Bauen.*

Baum, M. (2022). Die Sprache der Stadt–Architektur (-fotografie) zwischen ästhetischer Subjektivierung, Kommerzialisierung und normalisierten Blicken (Introduction), Ders. (Ed.), *Die Sprache der Stadt: Architektur-und urbane Raumbilder zwischen ästhetischer Subjektivierung und normalisierender Kommerzialisierung*, Schriften der Katholischen Hochschule Nordrhein-Westfalen, Opladen/Berlin/Toronto (pp. 11–32).

Baur, N., & Blasius, J. (Ed.) (2014). *Handbuch Methoden der empirischen Sozialforschung*, Wiesbaden.

Bechmann, G, & Frederichs, G. (2005). Problemorientierte Forschung: Zwischen Politik und Wissenschaft, *10 Years ITAS,* 43.

Beck, H. (2014). *Behavioral economics. Eine Einführung.* Wiesbaden.

Bhattacharya, S., & Chatterjee, A. (2022). Digital project driven supply chains: A new paradigm. *Supply Chain Management: An International Journal, 27*(2), 283–294.

Bisky, J. (2000). *Poesie der Baukunst: Architekturästhetik von Winckelmann bis Boisserée*, Heidelberg.

Blaschke, B. (2004). *Der homo oeconomicus und sein Kredit bei Musil, Joyce, Svevo, Unamuno und Céline*, Paderborn.

Blumenberg, H. (2020 [1986]). *Die Lesbarkeit der Welt*, Berlin.

Böhme, G. (1995). *Atmosphäre: Essays zur neuen Ästhetik*, Berlin.

Böhme, G. (2013). Atmosphären als Gegenstand der Architektur. In Baumberger, C. (Ed.), *Architekturphilosophie*, Kunstphilosophie (Vol. 10, pp. 94–110), Paderborn.

Böhme, G. (2020). Atmosphäre (sensu Gernot Böhme). In T. Kirchhoff (Ed.), *Online Encyclopedia Philosophy of Nature/Online Lexikon Naturphilosophie*, o. O.

Braun, D. (1993). *Zur Steuerbarkeit funktionaler Teilsysteme: Akteurtheoretische Sichtweisen funktionaler Differenzierung moderner Gesellschaften*, Wiesbaden.

Bullen, P., & Love, P. (2011). Factors influencing the adaptive re-use of buildings. *Journal of Engineering, Design and Technology.*

Burzan, N. (2005). *Quantitative Methoden der Kulturwissenschaften: Eine Einführung*, Stuttgart.

Calliess, C. (2021). Das „Klimaurteil“ des Bundesverfassungsgerichts: „Versubjektivierung“ des Art. 20 a GG. *Zeitschrift für Umweltrecht, 32*(6), 355–358.

Chase, J. (1991). The role of consumerism in American architecture. *Journal of Architectural Education, 44*(4), 211–224.

Chen, L., et al. (2023). Green construction for low-carbon cities: A review. *Environmental Chemistry Letters*, 1–31.

CO_2 Earth. (2023). Daily CO_2. Retrieved February 13, 2023, from https://www.co2.earth/daily-co2

Colgan, J. D., & Keohane, R. O. (2017). The liberal order is rigged: Fix it now or watch it wither. *Foreign Affairs, 96*(3), 36–44.

Cortes Vargas, T. C., et al. (2021). Carbon Conscious! The impact of embodied emissions on design decisions for building envelopes. In *Proceedings of the 9th PowerSKIN Conference*, TU Delft Open.

Dauermann, A., & Enneking, U. (2019). Die landwirtschaftliche Baukultur nachhaltiger gestalten und die kommunikative Wirkung der Bausubstanz erkennen. *Innovatives Denken für eine nachhaltige Land- und Ernährungswirtschaft, Beiträge zur 15. Wissenschaftstagung Ökologischer Landbau,* Kassel.

De Angelis, R., et al. (2018). Supply chain management and the circular economy: Towards the circular supply chain. *Production Planning & Control, 29*(6), 425–437.

Distelhorst, L. (2014). *Leistung: Das Endstadium der Ideologie*, Bielefeld.

Distelhorst, L. (2019). *Kritik des Postfaktischen: Der Kapitalismus und seine Spätfolgen*, Paderborn.

Dixit, M. K., et al. (2010). Identification of parameters for embodied energy measurement: A literature review. *Energy and Buildings, 42*(8), 1238–1247.

Dröge, S. (2015). *Das Pariser Abkommen 2015: Weichenstellung für das Klimaregime.* SWP-Studie.

Fehr, L. (2021). Auf dem Holzweg? Gestaltung neuer Wirtschafts- und Lebensweisen von Bioökonomie. *Hamburger Journal Für Kulturanthropologie, 13*, 224–234.

Ferdous, W., et al. (2021). Recycling of landfill wastes (tyres, plastics and glass) in construction—a review on global waste generation, performance, application and future opportunities. *Resources, Conservation and Recycling,* 173.

Foucault, M. (2002). *Schriften in vier Bänden*, Dits et écrits, Frankfurt/M. (p. 779).

Fraunholz, U., & Wölfel, S. (Eds.). (2012). Ingenieure in der technokratischen Hochmoderne, Thomas Hänseroth zum 60. Geburtstag. *Cottbuser Studien zur Geschichte von Technik, Arbeit und Umwelt,* 40, Münster.

Friesacher, H. (2022). Neue Technologien im Blick des Anerkennungs- und des Resonanzparadigmas. In M. Hülsken-Giesler, et al. (Eds.), *Neue Technologien für die Pflege: Grundlegende Reflexionen und pragmatische Befunde*, Göttingen (pp. 57–78).

Fuchs, T. (2020). *Verteidigung des Menschen: Grundfragen einer verkörperten Anthropologie*, Berlin.

Fuchs, T. (2021). *Das Gehirn—Ein Beziehungsorgan: Eine phänomenologisch-ökologische Konzeption*, Stuttgart.

Fukuyama, F. (2022). *Liberalism and its Discontents*, London.

Furrer, B. (1992). Agrikultur-Baukultur. *Baukultur—Wohnkultur—Ökologie: Tagungsband zum 5. interdisziplinären Symposium an der Universität Zürich im April 1992.* Wiesbaden.

Gates, B. (2021). Wie wird die Klimakatastrophe verhindert, Munich. (Or also here online: Retrieved February 24, 2023, from https://www.gatesnotes.com/Concrete-in-China

Geertz, C. (2000). *The interpretation of cultures*, New York.

Gendlin, E., & Wiltschko, J. (1999). Focusing *in der Praxis Eine schulenübergreifende Methode für Psychotherapie und Alltag,* Stuttgart.

Georgescu-Roegen, N. (1975). Energy and economic myths. *Southern Economic Journal, 41*, 347–381.

Gerrig, R. J., & Zimbardo, P. G. (2008). *Psychologie*, Munich.

Giurca, A. (2022). Why is communicating the circular bioeconomy so challenging? *Circular Economy and Sustainability*, 1–10.
Gkountani, V. A., & Tsoulfas, G. T. (2023). Circular bioeconomy: A review on the current state and future opportunities. In *Digital economy and the green revolution: 16th international conference on business excellence, ICBE 2022 in Bucharest, Cham.*
Glasze, G., & Mattissek, A. (Eds.). (2021). *Handbuch Diskurs und Raum: Theorien und Methoden für die Humangeographie sowie die sozial- und kulturwissenschaftliche Raumforschung*, 11, Bielefeld.
Glock, C., et al. (2022). Massivbau in Zeiten von Klimawandel und Ressourcenverknappung—Herausforderungen und Lösungsansätze/Concrete construction in times of climate change and resource shortage—challenges and solutions. *Civil Engineer, 97*(1–2), 1–12.
Godemann, J., & Bartelmeß, T. (2021). *Ernährungskommunikation: Interdisziplinäre Perspektiven—Theorien–Methoden*, Wiesbaden.
Godfrey, D. M., & Feng, P. (2017). Communicating sustainability: Student perceptions of a behavior change campaign. *International Journal of Sustainability in Higher Education.*
Goh, C. S., et al. (2020). Qualitative study of sustainability policies and guidelines in the built environment. *Journal of Legal Affairs and Dispute Resolution in Engineering and Construction, 12*(2).
Görgl, P. J. (2008). *Die Amerikanisierung der Wiener Suburbia?: Der Wohnpark Fontana. Eine sozialgeographische Studie*, Wiesbaden.
Goubran, S. (2019). On the role of construction in achieving the SDGs. *Journal of Sustainability Research, 1*(2).
Gould, H., et al. (2023). Trends and policy in bioeconomy literature: A bibliometric review. *EFB Bioeconomy Journal.*
Grandits, D., & von Planta, C. (2022). High-quality Baukultur of the past'–building research is needed: Building research as sine qua non for the 'high-quality Baukultur of the past. *Monumenta, 5*, 90–93.
Grazuleviciute-Vileniske, I., et al. (2021). The role of aesthetics in building sustainability assessment. *Spatium*, 79–89.
Grossarth, J. (2018). *Die Vergiftung der Erde. Metaphern und Symbole agrarpolitischer Diskurse seit Beginn der Industrialisierung*, Frankfurt/New York.
Grossauer, F., & Stoeglehner, G. (2023). Bioeconomy—a systematic literature review on spatial aspects and a call for a new research agenda. *Land, 12*(1), 234.
Grunwald, A. (2014). *Technikzukünfte als Medium von Zukunftsdebatten und Technikgestaltung.* KIT Scientific Publishing.
Grunwald, A. (2020). The objects of technology assessment. Hermeneutic extension of consequentialist reasoning. *Journal of Responsible Innovation, 7*(1), 96–112.
Guidetti, E., & Ferrara, M. (2023). Embodied energy in existing buildings as a tool for sustainable intervention on urban heritage. *Sustainable Cities and Society,* 88.
Gupta, H. (2020). Barriers and overcoming strategies to supply chain sustainability innovation. *Resources, Conservation and Recycling,* 161.
Gursel, I., et al. (2009). Modeling and visualization of lifecycle building performance assessment. *Advanced Engineering Informatics, 23*(4), 396–417.
Habermas, T., & Köber, C. (2015). Autobiographical reasoning in life narratives buffers the effect of biographical disruptions on the sense of self-continuity. *Memory, 23*(5), 664–674.
Halofsky, J. E., et al. (2020). Changing wildfire, changing forests: The effects of climate change on fire regimes and vegetation in the Pacific Northwest, USA. *Fire Ecology, 16*(1), 1–26.
Hempel, C., et al. (2019). Bioökonomie aus Sicht der Bevölkerung, *Thünen Working Paper*, 115.
Hertwich, E. G., et al. (2019). Material efficiency strategies to reduce greenhouse gas emissions associated with buildings, vehicles, and electronics—a review. *Environmental Research Letters, 14*, 4.
Hirschfelder, G. (2012a). Esskultur: Ernährungsalltag zwischen Ideal Trend und Realität. *Ernährung & Medizin, 27*(03), 107–110.

Hirschfelder, G. (2012). Europäischer Alltag im Fokus der Kulturanthropologie/Volkskunde. In S. Conermann (Ed.), *Was ist Kulturwissenschaft?—Zehn Antworten aus den „kleinen Fächern“* Bielefeld (p. 152).

Hirschfelder, G., & Pollmer. P. (2020). Das Wissen vom „guten“ Essen—Ernährungskommunikation als historisches Paradigma. In J. Godemann & T. Bartelmeß (Eds.), *Ernährungskommunikation: Interdisziplinäre Perspektiven–Kontexte–Methodische Ansätze* (pp. 61–73).

Hitzler, S. (2020). Ernährung als Alltagsdilemma–Konversationsanalytische Einblicke in eine Selbsthilfegruppe von Personen mit Übergewicht. In J. Godemann & T. Bartelmeß (Eds.), *Ernährungskommunikation: Interdisziplinäre Perspektiven–Kontexte–Methodische Ansätze* (pp. 1–16).

Hodgson, G. M. (2004). *The evolution of institutional economics*, London.

Hollberg, A. (Ed.) (2020). Evaluation of BIM-based LCA results for building design. *Automation in Construction,* 109.

Hosey, L. (2012). *The shape of green: Aesthetics, ecology, and design*, Washington.

Hossain, U., et al. (2020). Circular economy and the construction industry: Existing trends, challenges and prospective framework for sustainable construction. *Renewable and Sustainable Energy Reviews,* 130.

Irrgang, B. (2010). *Homo Faber: Arbeit, technische Lebensform und menschlicher Leib*, Würzburg.

Karlsson, I., et al. (2021). Achieving net-zero carbon emissions in construction supply chains–a multidimensional analysis of residential building systems. *Developments in the Built Environment,* 8.

Kelle, U. (2022). Mixed methods. In *Handbuch Methoden der empirischen Sozialforschung*, Wiesbaden (pp. 163–177).

Knoblauch, H. (2016). *Die kommunikative Konstruktion der Wirklichkeit*, Wiesbaden.

Krickhahn, T. (2022). Bioökonomie—Eine (inter-)disziplinäre Perspektive. In H. Jeschke (Ed.), *Bioökonomie: Impulse für ein zirkuläres Wirtschaften*, Wiesbaden (pp. 21–36).

Kriese, U., & Scholz, R. W. (2012). Lifestyle ideas of house builders and housing investors. *Housing, Theory and Society, 29*(3), 288–320.

Kurt, R. (2008). Vom Sinn des Sehens. Phänomenologie und Hermeneutik als Methoden visueller Erkenntnis. In J. Raab, et al. (Eds.), *Phänomenologie und Soziologie: Theoretische Positionen, aktuelle Problemfelder und empirische Umsetzungen*, Wiesbaden (pp. 369–378).

Kurt, R., & Herbrik, R. (2014). Sozialwissenschaftliche Hermeneutik und hermeneutische Wissenssoziologie. In N. Baur & J. Blasius (Eds.), *Handbuch Methoden der empirischen Sozialforschung*, Wiesbaden.

Kussin, M., & Berstermann, J. (2022). Maßnahmen und Instrumente. In This (Eds.), *Agrarkommunikation: Eine Einführung in Theorie, Konzeption und Umsetzung*. Wiesbaden (pp. 91–135).

Lang, D. J., et al. (2012). Transdisciplinary research in sustainability science: Practice, principles, and challenges. *Sustainability Science, 7*, 25–43.

Lavdas, A., & Salingaros, N. (2022). Architectural beauty: Developing a measurable and objective scale. *Challenges, 13*, 2.

Lewandowski, I., et al. (2000). Miscanthus: European experience with a novel energy crop. *Biomass and Bioenergy, 19*(4), 209–227.

Liessmann, K. P. (2009). *Schönheit*, Vienna.

Lima, L., et al. (2021). Sustainability in the construction industry: A systematic review of the literature. *Journal of Cleaner Production,* 289.

Luhmann, N. (1984). Die Wirtschaft der Gesellschaft als autopoietisches System. *Zeitschrift Für Soziologie, 13*(4), 308–327.

Luhmann, N., et al. (1990). „Zunächst einmal schlage ich gar nichts vor...“, Interview with Niklas Luhmann, Part II: Über ökologische Kommunikation und Moral. *Zeitschrift für Entwicklungspädagogik, 13*(3), 24–28.

Luhmann, N., & Heidbrink, L. (2004). *Ökologische Kommunikation*, Wiesbaden.

Lützkendorf, T., & Balouktsi, M. (2022). Embodied carbon emissions in buildings: Explanations, interpretations, recommendations. *Buildings and Cities, 3*, 1.

Martin, H., et al. (2023). Reconsidering actor roles in regional innovation systems: Transformative industrial change in the forest-based bioeconomy, *Regional Studies*, 1–13.
Marzouk, M., & Elshaboury, N. (2022). Science mapping analysis of embodied energy in the construction industry. *Energy Reports, 8*, 1362–1376.
McKenzie-Mohr, D. (2011). *Fostering sustainable behavior: An introduction to community-based social marketing*. New Society Publishers.
McMahon, J. A. (2005). Beauty. In *The Routledge companion to aesthetics*. London (pp. 245–256).
Meier-Seethaler, C. (2001). *Gefühl und Urteilskraft: Ein Plädoyer für die emotionale Vernunft*, Munich.
Meyer, E. K. (2008). Sustaining beauty. The performance of appearance: A manifesto in three parts. *Journal of Landscape Architecture, 3*(1), 6–23.
Mishra, A., et al. (2022). Land use change and carbon emissions of a transformation to timber cities. *Nature Communications, 13*, 1.
Mora, C., et al. (2018). Broad threat to humanity from cumulative climate hazards intensified by greenhouse gas emissions. *Nature Climate Change, 8*(12), 1062–1071.
Motschnig, R., & Nykl, L. (2009). *Konstruktive Kommunikation: Sich und andere verstehen durch personenzentrierte Interaktion*, Stuttgart.
Müller, E., & Schmieder F. (2016). *Begriffsgeschichte und historische Semantik. Ein kritisches Kompendium*, Berlin.
Munaro, M. R. (2020). Towards circular and more sustainable buildings: A systematic literature review on the circular economy in the built environment. *Journal of Cleaner Production,* 260.
Nassehi, A. (2021). Trägheit der Transformation (Titel der Mitschrift), Vortrag gehalten auf dem Forum Zukunft Bau, Bonn, Video documentation. Retrieved February 13, 2023, from https://www.zukunftbau.de/veranstaltungen/zukunft-bau-kongresse/2021/videodokumentation
Nguyen, A. T. (2019). The relationship between economic growth, energy consumption and carbon dioxide emissions: Evidence from central Asia. *Eurasian Journal of Business and Economics, 12*(24), 1–15.
Niklas, K. J., & Kerchner, V. (1984). Mechanical and photosynthetic constraints on the evolution of plant shape. *Paleobiology, 10*(1), 79–101.
Paddock, T. (2004). Gedachtes Wohnen: Heidegger and cultural geography. *Philosophy & Geography, 7*(2), 237–249.
Pelicice, F. M., & Castello, L. (2021). A political tsunami hits Amazon conservation. *Aquatic Conservation: Marine and Freshwater Ecosystems, 31*(5), 1221–1229.
Pierskalla, C. D., et al. (2016). Examining the product and process of scenic beauty evaluations using moment-to-moment data and GIS: The case of Savannah, GA. *Urban Forestry & Urban Greening, 19*, 212–222.
Pinsdorf, C. (2022). Bioeconomy beneath and beyond: Persisting challenges from a philosophical and ethical perspective. In: *Bioeconomy and sustainability: Perspectives from natural and social sciences, economics and ethics*, Cham (pp. 343–377).
Pons. (2023). Colere. Retrieved January 12, 2023, from https://de.pons.com/%C3%BCbersetzung/latein-deutsch/colere
Pyka, A. (2017). Die Bioökonomie unter dem Blickwinkel der Innovationsökonomiein. In Pietzsch, J. (Ed.), *Bioökonomie für Einsteiger* (pp. 129–138).
Qu, K., & Liu, Z. (2022). Green innovations, supply chain integration and green information system: A model of moderation. *Journal of Cleaner Production,* 339.
Radkau, J. (2011). *Die Ära der Ökologie: Eine Weltgeschichte*, Munich.
Rebolledo-Leiva, R., et al. (2023). Progress of social assessment in the framework of bioeconomy under a life cycle perspective. *Renewable and Sustainable Energy Reviews,* 175.
Rittel, H. W., & Webber, M. (1974). Wicked problems. *Man-Made Futures, 26*(1), 272–280.
Röck, M., et al. (2020). Embodied GHG emissions of buildings—the hidden challenge for effective climate change mitigation. *Applied Energy,* 258.
Rosa, H. (2016). *Resonanz*, Berlin.

Sánchez-Garrido, A. J., et al. (2022). Multi-criteria decision-making applied to the sustainability of building structures based on Modern Methods of Construction. *Journal of Cleaner Production*, 330.
Schäfer, M., et al. (2012). Life events as windows of opportunity for changing towards sustainable consumption patterns? Results from an intervention study. *Journal of Consumer Policy, 35*, 65–84.
Scheer, D., & Konrad, W. (2020). Der Zukunftsdiskurs Bioökonomie im Spiegel der gegenwärtigen Themenlandschaft. *Bioökonomie nachhaltig gestalten: Perspektiven für ein zukunftsfähiges Wirtschaften, Technikzukünfte Wissenschaft und Gesellschaft/Futures of Technology, Science and Society*, 43–69.
Schellnhuber, H.-J. (2021). Lecture given at the Forum Zukunft Construction, Bonn. Retrieved March 23, 2023, from https://www.zukunftbau.de/veranstaltungen/zukunft-bau-kongresse/2021/videodokumentation
Schellnhuber, H. J. (2024). Bauhaus earth: Sustainable use of wood in the construction sector. In *3 Degrees More: The impending hot season and how nature can help us prevent it*, Cham (pp. 147–177).
Schleissing, S. (2022). Bioökonomie als Herausforderung für eine evangelische Ethik. *Zeitschrift für Evangelische Ethik, 66*(3), 198–211.
Schlicht, E. (1998). *On custom in the economy*, Oxford.
Schmidlehner, M. (2023). The green growth narrative, bioeconomy and the foreclosure of nature. *Sustainability Science*, 1–14.
Schnell, D. (2022). How should we interpret the Davos Declaration? *Monumenta, 5*, 25–27.
Schöggl, J.-P., et al. (2020). The narrative of sustainability and circular economy—a longitudinal review of two decades of research. *Resources, Conservation and Recycling*, 163.
Schummer, J. (2006). *Symmetrie und Schönheit in Kunst und Wissenschaft, Ästhetik in der Wissenschaft* (pp. 59–78).
Schweppenhäuser, G. (2019). Kulturindustrie, *Handbuch Kritische Theorie* (pp. 1079–1104).
Sev, A. (2009). How can the construction industry contribute to sustainable development? A conceptual framework. *Sustainable Development, 17*(3), 161–173.
Shukla, P. R., et al. (2019). Climate Change and Land: An IPCC special report on climate change, desertification, land degradation, sustainable land management, food security, and greenhouse gas fluxes in terrestrial ecosystems, n.p.
Siegel, K. M., et al. (2020). When international sustainability frameworks encounter domestic politics: The sustainable development goals and agri-food governance in South America. *World Development*, 135.
Skillington, K., et al. (2022). A review of existing policy for reducing embodied energy and greenhouse gas emissions of buildings. *Energy Policy*, 168.
Spiller, A., & Nitzko, S. (2021). *Ernährung und Gesundheit: Forschungsansätze und Diskurse der Ernährungspolitik*, Wiesbaden.
Stein, M., et al. (2018). Wir sind die Bioökonomie „– Perspektiven von Akteuren aus dem deutschen Forst- und Holzsektor." *Allgemeine Forst-und Jagdzeitung, 189*, 30–40.
Tateo, L. (2017). *Giambattista Vico and the new psychological science*, London.
Thevs, N., et al. (2022). Potential of agroforestry to provide wood resources to central Asia. *Forests, 13*, 8.
Tomor, Z., et al. (2019). Smart governance for sustainable cities: Findings from a systematic literature review. *Journal of urban technology, 26*(4), 3–27.
Traue, B., et al. (2022). Diskursanalyse, Handbuch Methoden der empirischen Sozialforschung. In: N. Baur & J. Blasius (Eds.), *Handbuch Methoden der empirischen Sozialforschung*, Wiesbaden (pp. 627–645).
Triantafyllidis, T. (2001). Eine Brücke zwischen Empirie und Wissen. *Bautechnik, 78*(4), 235–238.
Tveit, M., et al. (2018). Scenic beauty: Visual landscape assessment and human landscape perception. *Environmental psychology: An introduction*, 45–54.
Uekötter, F. (2012). Eine ökologische Ära? Perspektiven einer neuen Geschichte der Umweltbewegungen. *Zeithistorische Forschungen - Studies in Contemporary History, 9*(1), 108–114.

UN. (2022). Demographic Profiles. Retrieved February 24, 2023, from https://population.un.org/wpp/Graphs/DemographicProfiles/Line/900
Urmetzer, S., et al. (2020). Learning to change: Transformative knowledge for building a sustainable bioeconomy. *Ecological Economics,* 167.
Van der Loo, H., & van Reijen, W. (1992). *Modernisierung, Projekt und Paradox*, Munich.
Vandenbroele, J., et al. (2020). Nudging to get our food choices on a sustainable track. *Proceedings of the Nutrition Society, 79*(1), 133–146.
Vilsmaier, U., & Lang, J. (2014). Transdisziplinäre Forschung. *Nachhaltigkeitswissenschaften*, 87–113.
Vogelpohl, T., & Töller, A. (2021). Perspectives on the bioeconomy as an emerging policy field. *Journal of Environmental Policy & Planning, 23*(2), 143–151.
Vogt, M. (2018). Bedingungen ethisch verantwortbarer Bioökonomie. *Forum Wirtschaftsethik*, 26.
Wagner, B., et al. (2022). *Erfolgspotenziale der Holzhybridbauweise in der Immobilienprojektentwicklung*, Wiesbaden.
Warda, J. (2020). Wohnen im Einfamilienhaus. In F. Eckardt & S. Meier (Eds.), *Handbuch Wohnsoziologie*, Wiesbaden (pp. 1–20).
Watzlawick, P. (2015). *Man kann nicht nicht kommunizieren: Das Lesebuch*, Zurich.
WGBC. (2023). World Green Building Council Commitment. Retrieved February 13, 2023, from https://worldgbc.org/thecommitment/commitment-signatories/?_sfm_signatory_type=cities
Wiltschko, J. (2000). Felt Sense. In *Wörterbuch der Psychotherapie* (pp. 205–206).
Winkler, B., et al. (2019). Urban gardening in Germany: Cultivating a sustainable lifestyle for the societal transition to a bioeconomy. *Sustainability, 11*, 3.
Wittmayer, J. M., et al. (2019). Narratives of change: How social innovation initiatives construct societal transformation. *Futures,* 112.
Zhuang, W., et al. (2021). On the factors influencing green purchase intention: A meta-analysis approach. *Frontiers in Psychology,* 12.

Bioeconomy: Perspectives, Traditions, and Tensions

3

Abstract

This chapter explores the diverse approaches that make up the bioeconomy as an interdisciplinary field. One major approach focuses on agriculture and forestry, aiming to substitute resources sustainably. The other, a broader political-economic strategy, seeks a fundamental transformation of economic systems. Both are well-rooted in the EU, and numerous other countries worldwide. But the bioeconomy discourse goes beyond merely channeling attention toward material flows in economics and politics. The metaphor itself is powerful: “bios” meaning “life” evokes a vision of the bioeconomy as an alternative to a tech-driven, mechanistic society. In this way, the bioeconomy is positioned as a corrective, offering a balanced, life-centered view. This chapter traces the evolution of these two primary visions within bioeconomy and highlights the pivotal role of innovation—both as technical progress and as socio-cultural transformation.

What can be the benefit of speaking about bioeconomy and construction at the same time? What is the bioeconomy of construction? The introduction and Chap. 2 have attempted an introduction at the levels of facts and significance. The following approach the bioeconomy as it is defined, institutionally shaped, and historically evolved.

The chapter begins with references to sustainability policy, followed by an essayistic exploration of an architectural-cultural concept of the bioeconomy of buildings. It identifies a bioeconomic tradition of language that is associative, concrete, and scientific-abstract simultaneously, as seen in the work of Nicholas Georgescu-Roegen, the ‘academic founding father’ of economic bioeconomics. Next, the chapter examines various disciplinary and practical approaches to the field of bioeconomy, and then highlights the increasing intersectoral material connectedness resulting from bioeconomic developments. This is followed by

J. Grossarth, *Bioeconomy of Buildings*, https://doi.org/10.1007/978-3-031-84014-2_3

overviews of the relevance of the concept of bioeconomy in politics and the common definitions of bioeconomy. Two traditions of bioeconomy are then elaborated in more detail. These *two cultures* of bioeconomy are referred to as *resource-oriented* and *political-economic.* After a brief discussion of various views on whether the bioeconomic approaches imply a statement about the possibility of sustainable economic growth, the suggestion follows to understand the bioeconomy beyond the usual paths also as a paradigm for historical foresight and cultural-critical reflection that resonates in politics, science and society.

3.1 Bioeconomy in Construction Within Sustainability Policy

In 2019, the European Union published the "New Green Deal" as an industrial policy strategy—renamed in 2025 to "Clean Industrial Deal"—which can be described as a consistent alignment of economic and research policy towards a circular economy, carbon neutrality and economic growth (Bonoli et al., 2021). In the beginning of the 2020s, climate protection policy has been the main area of government action in large parts of the world (Chap. 2).

3.1.1 Greenhouse Gas Reduction and the Bioeconomy

The greenhouse gas equivalents that need to be saved by 2030 in the sectors "buildings" and "industry" relevant to construction are of substantial magnitude. The use of *biomass* as a climate-neutral or even climate-positive fuel, material or nutrient in all sectors plays a central role. In order to achieve resource-efficient use of biomass, the substances should be kept in the economic cycle as long as possible. The increasing incentives and obligations for life cycle assessment (*LCA*) with a view to greenhouse gas emissions are likely to profoundly change the construction sector in a few years.

The bioeconomy, with its focus on renewable resources, is therefore moving into focus in a new and different way. The expansion plans proposed by climate science are so large that they are reminiscent of the technocratic dimensions of *geoengineering* (Soeder, 2022): The study by Churkina et al. (2020), it is stated that the construction of wood for more than two billion new urban residents expected worldwide from 2020 to 2050 could store 0.01–0.68 Gt of carbon annually—depending on the forecast scenarios and assumed living space requirements. Over 30 years, the carbon emissions would have to be reduced by a total of 0.3–20 Gt. The study contains several calculation scenarios. Such a project could not only succeed through traditional timber construction. Instead, one should imagine a global pulp bioeconomy of construction: "Wood from existing and future buildings (which are designed for easy disassembly) must be recovered and reused as a raw material source for consumer products or the buildings of the coming generation" Churkina et al. (2020).

3.1.2 Opportunities and Risks of the Bioeconomy

The politically decided decarbonisation represents a turning point that can be described from different perspectives. From a materials science perspective, the use of metallic and mineral building materials relative to polymers, which also include cellulose, is likely to decrease. From an energy history perspective, a turning point could be imminent that makes energy more expensive, scarcer, and the embodied energy more valuable. From an agricultural, forestry and environmental history perspective, the pressure on the area is likely to increase if more wood and agricultural raw materials are to be used in construction, as well as in other sectors. At the same time, there is the historically new possibility of preventing determinisms of land scarcity through the expansion of renewable energy (Turnbull, 2021). Overall, the path of the bioeconomy offers opportunities and risks that can be outlined as follows (Table 3.1; Thrän & Moesenfechtel, 2020, p. 4f., additions JG).

3.2 The Potential of the Bioeconomy

When I tell people in my environment that I am a professor of bioeconomy, they often think of organic stores. They believe, as their looks or words reflect to me, that people like me are working to make the entire economy like an organic store: chemical-free, fragrant, expensive, exclusive. Although the supporting science of the bio-economy, based on natural science, has no such intentions. But the association that the term obviously evokes among less informed *laypeople* is to be taken seriously.

The bioeconomy has been shaped in recent decades by political practice—but also, independently of this, by academic writings, which were located in many different academic disciplines. To understand this history of the term is a necessary and enlightening task. Similar to its sister, sustainability, the bioeconomy is suitable for appropriations.

Table 3.1 Opportunities and risks of the bioeconomy

Opportunities	Risks
• Secure food supply for the world's population through further development of production systems (new proteins, aquaculture, urban farming …) • Substitution of fossil commodities • Improved material properties, smarter processes and products • More durable construction, construction as a CO2 storage • New jobs in rural areas	• Increasing conflicts over biomass and land use competitions • Ethically controversial questions: genetic engineering, agricultural-forestry intensification • "Economisation" of the living (Gottwald & Krätzer, 2014) • Bureaucratic, "eco-planned economic" overburdening of entrepreneurial actors through overambitious goals, such as the reduction of greenhouse gases

Why not start thinking about the bioeconomy at the starting point of the association that the word triggers? That would be the *organic store economy*. This metaphor may contain atmospherically pleasant aspects, but it also harbors economic-social explosive power (there would be winners and losers; the organic store is not known for cheap prices). Such considerations unfold when the bioeconomy is subsumed under the umbrella of sustainability.

But that's not what this introduction is about. It's about the images associated with the *bios-economy*—and their impact, their bodily empathy (*embodiment*).

The attractiveness of a *feeling of bioeconomic building*—an experience-based, bodily comprehensible understanding—may explain the success of non-fiction literature and YouTube films that deal with nature or nature-like living: trees, gardens, chickens. But also: half-timbered houses, clay plaster, tiny house. The "biopoetics" inherent in these words and images (Andreas Weber) probably also lies in the fact that they contain an implicit promise of resistance. In addition, it opens up the space of inspiration for engineers in biology—organic forms, lightweight support structures, green roofs and facades with versatile use (Fig. 3.1).

Fig. 3.1 The green roof of the Anders Hus in Odense (2022). Green roof in Darmstadt (2023). Tiny Houses in the form of trailers in the ecovillage Siebenlinden (2010). Photos by Jan Grossarth

3.2.1 Home as Cultural Identity: Insights from Wendell Berry

What associative spaces does the word "bio-economy" open? Imaginations of a happy life, associated with values such as neighborhood, care, responsibility in the immediate vicinity of people? *Bio* would then also mean joy of life. The connection of work and meaning, sense-making, of economy with community reference and creativity would then be crucial. The separation of rationalistic Professional logics and emotion-related leisure would be highly problematic. Are these levels also part of the sustainability problem?

The "Cartesian" split into objective reason of science and unscientific subjectivity is described by philosophers, psychologists and psychiatrists as tragic. Authors of these sciences have also become aware again of the pre-rational of bodily sensation (Sect. 2.3.5). References to the tragic split go back to Giambattista Vico (1668–1744) and are brought to the present (Pern, 2019). Vico associates creativity, fantasy, signification, meaning attribution with the memory of the human mind and its "poetic structure" (Pern, 2019, p. 474). What derivation could be made from these thoughts with regard to the complex of topics of sustainable building—especially with regard to the high degree of division of labor and specialisation? Put differently: What does the effect of desolate functionality of a gray "passive-house" with gravel front garden testify in this sense? Does such a building represent a good worth preserving for future generations?

When it comes to the broad semantic field of bioeconomy, an excursion into the field of agriculture is certainly appropriate. The writer Wendell Berry has given the debate about organic, sustainable agriculture an original poetic imprint. When he was invited to give a lecture at an agricultural university in 1974, he expressed distance from this institution by opening his speech with his bewilderment with the lecture title. It had been pre-formulated for him by the agricultural engineers:

> I was asked to talk about 'labor intensive microsystems agriculture'. That's not my language. It's not the sort of language that I wish to use, because it's the way people speak when they don't want be understood by most people. (Berry, 1974, as well as following quotes)

Berry then went on to explain that academic expressions like "labor-intensive, small-scale agriculture" overemphasized structural or technological aspects of agriculture. This was misleading. He explicitly distanced himself from the academic approach:

> Perhaps I say something about my qualification. I'm not a farm expert, and I wasn't educated in an agricultural college. I do come from a farm community, where my people have farmed five or six generations before me., and where I farm, and inherited my father's and others' concerns what happened to farmers.

Later, he spoke of memories of gardens as well as buildings:

> They grew gardens. They produced their own meat, milk and eggs. They were hardly diversified. (…) The proud of most people were still at their homes. And their homes looked like it.

Anticipating the obvious accusation of romanticisation, or even technophobia, he then added:

> This was my no means a perfect society. (…) But I have spoken of its agricultural economy, of a generation ago, to suggest that there has also been good qualities and dignities to it, that might have been cultivated and build upon. But they were not cultivated and built upon. But they were repudiated as the stuff of a hopeless outlawed unscientific way of life.

Instead of an evolution of agriculture—and construction—technological progress brought radical cultural breaks. This affects not only farming methods, but also the disappearance of livestock markets in rural areas, of peasant marriage strategies and the self-understanding of children (Frie, 2023).

Berry's crucial point is: It's not the technology that's the problem, but certainly: the arrogance of the technical sciences and the engineers. This arrogance is a consequence of technical and institutionalized power and the monetary strength that technical development gives to people. But arrogance is also a vehicle of power. In the name of progress, the "good qualities and dignities" of the old craft are no longer seen as a concern. Moreover, those who remind us are mocked.

Wendell Berry refers to the sociocultural, and thus inseparably linked, aesthetic qualities of *community, cultural integrity* and a *craft pride*. He has also identified the "unscientific" as a core accusation with which a technical-academic expertocracy knows how to distinguish itself from such concerns. And in this sense, shouldn't the relationship of some civil engineers, building material manufacturers or builders to the people who advocate for concerns of building culture also be critically examined?

The beneficiaries of progress gain land, machines, power and professorships. The left-behinds suffer not only from loss, but also from the habitus of the winners of this one-sided history of technology. For the tragedy of loss, says the Catholic Berry, is assigned to the private sphere of those affected—as refusal of education, eccentricity or personal failure. For him, this, too, is the monstrous irresponsibility of technical expert power:

> And it's a work of monsters' ignorance and irresponsibility on the part of the experts and politicians who have prescribed, encouraged and applauded the disintegration of such farming communities all over this country into our miraculous modern American agriculture.

What is called "progress" is in truth a disintegration and displacement of millions of people from their places of action and tradition, Wendell Berry then states. The paradigm of "*get big or get out*" would result in comparable cultural revolutions and human impoverishments like the agrarian reforms in communism, says Berry. He substantiates this thesis as follows:

> The only difference here is in method. The first use for the communists is military. With us, it is mainly economy. A free market in which the freest where the richest. The attitudes were equally cruel. And I believe that in the long run the results will be equally damaging. (…) A healthy culture is a communal order of memory, insight, value and aspiration.

If it is academically and especially politically difficult to name universally valid qualities of the "good life", Berry does not make it too complicated. He speaks of a "healthy culture". And this, he says, is:

- community-oriented
- history-related
- rich in experience
- value-bound and
- hopeful.

Technology would be very welcome for the facilitation of work. But it is not the source of a passionate relationship to work but can also be an obstacle for the reasons indicated. It can become a means to an end. And that is why Wendell Berry looks with skepticism at the effects of a (technical) scientific education. He says professional passion is rarely based on academic education:

> But this attitude does not come from technique or technology. It does not come from education. In more than two decades in universities, I have rarely seen it. It does not come even from principle. It comes from a passion that is culturally prepared. A passion for excellence and order, that is characteristically and maybe exclusively handed down to young people (…)

The "passion that is culturally prepared" and as a privilege of youth is handed down like a gift by (love-connected) elders, would thus be the core of a "good economy."

Are there approaches, *Bio-economy as a metaphor* so simple, so radically simple to understand? Doesn't everything Wendell Berry says also apply to the construction industry? Wouldn't "bios", life, in a culturally broad understanding of the bioeconomy, also be understood in the resonance space of such qualities of life? How do children actually paint ideal(ized) houses (Fig. 3.2)? The hermeneutics of children's drawings is—by the way—a worthwhile research subject (Daum, 2019).

What is the use of referring to Berry in the 2020s—half a century after this speech? The qualities of life he mentioned would not simply have to be defended against the encroachments, but rediscovered, against established routines and powerful interests. Failure is highly likely. However, to fall silent in the face of this insight only leads to resignation or cynicism. Berry opts for remembrance and narration, and this should not be confused with romanticisation or nostalgia. As the digital revolutions of artificial intelligence (AI) fuel concerns about a loss of individuality in and machine-based decisions in the 21th century, hermeneutics of "Berryesque cultural criticism" brings the individual perspective back into debate. By referring to it, however, it does not deny the social necessity of role attribution

Fig. 3.2 The poetry of building: Child's drawing of a country house. Illustration Frieda Theresia Grossarth

in modern societies. That is ambivalent: Role attributions come with impositions and expectations of conformity. Where the individual history does not fully align with the expectations of others, spaces for experiences of difference emerge. These can be painful but also productive for shaping one's life. Nevertheless, such cultural criticism should be taken seriously, but also critically reflected upon regarding its suitability as a political compass in a pluralistic and already highly modern society.

3.2.2 Georgescu-Roegen's Bioeconomy

For Georgescu-Roegen, the bioeconomy is actually the almost utopian grand metaphor for a permanently stable form of life and economy. He oscillates in his writings remarkably between mathematical proof in the model—which he uses against the neoclassical economies—and highly associative embeddings. The following doomsday fiction is an example of this:

> Perhaps, the destiny of man is to have a short, but fiery, exciting and extravagant life rather than a long, uneventful and vegetative existence. Let other species – the amoebas, for example – which have no spiritual ambitions inherit an earth still bathed in plenty of sunshine.

This passage is an example of unashamedly speculative, prophetic language. On the other hand, it reveals a latent ironic and poetic style: The amoebas, ambitionless, bask in eternal sunshine, while the extravagant humanity was all too short-lived! Georgescu-Roegen does not separate formal science from an interest in fate, event, human tragedy. Georgescu-Rogen's roles are those of the seer, admonisher, intellectual provocateur. His language, his concerns of technology-related progress skepticism in connection to *environmental destruction* and *resource depletion*, indicate: An understanding of bioeconomy, that does not consider the qualities of "Bios" in a broad horizon, would certainly not be one in the tradition of its founder. He does not hide his interpretations and his perspective behind formulas but uses them as a tool.

3.3 Defining the Bioeconomy: Two Traditions

The term (circular) bioeconomy is associated with comprehensive ideas of a sustainable economy as a whole. On the one hand, fossil resources should be replaced by renewable raw materials and renewable energy sources. On the other hand, the Material flows in the economy and in international trade should be as closed as possible. Both approaches were not invented with climate protection policy. Both traditions of thought have—different from each other—origins that touch and overlap each other in current political approaches (Grossarth, 2022; Wei et al., 2022).

The following proposes a conceptual distinction into two "schools" or traditions of bioeconomy. In political practice or in the use of terms, these may not be separated from each other. The proposed distinction is ideal–typical (Fig. 3.1). But it also aligns with the historical development of ideas and institutional evolution of the two bioeconomy traditions. The distinction offered here is into a

- **resource-oriented tradition** ("Resource Bioeconomy"), which considers agricultural-forestry and environmental engineering sectors under bioeconomy and organizes research funds and institutionalisation for this, associated with great hopes for innovation in biotechnology and a
- **political-economic tradition** of bioeconomy ("Political-economic Bioeconomy"), which ultimately wants to transform the entire economy towards the sustainability of material flows, and also wants to change the behavior of citizens and fiscal policy instruments towards this goal. Different ideas about the possibility of sustainable development can also be linked in the sense of economic growth (later subchapter).

This distinction is not established, but finds correspondences in the bioeconomic research literature, which still asks today what bioeconomy actually is. The dominance of natural and environmental engineering sciences is evident. An evaluation of 453 research papers up to 2016 showed that the most common fields of

	Bio-technology vision	Bio-resource vision	Bio-ecological vision
Goals	growth	growth sustainability	sustainability biodiversity ecosystem conservation
Value creation	biotechnology value creation through research and development	process-orientated: conversion and upcycling of bioresources	high-quality regional products integrated product systems
Innovation drivers	patents, founders, research and development	land use optimisation, interdisciplinarity, soil improvement, waste management	agroecological production, ethics, transdisciplinarity, circular value chains,

Selection JG; according to Bugge et al. 2016, p.10

Fig. 3.3 Three visions of the bioeconomy. JG

publication were biotechnology and applied microbiology, followed by energy economics, environmental sciences, chemistry, environmental engineering, and food sciences (Brugge et al., 2016, p. 8). In the search for the traditions of bioeconomy, the categorisation made on this basis is revealing. It distinguishes between a (1.) biotechnology vision, a (2.) bio-resource vision, and a (3.) bio-ecology vision (Fig. 3.3).

Especially the bio-ecology vision offers points of connection for the hermeneutic reading. In all cases, the bioeconomy approach promotes intersectoral connections, at least at the level of material flows (biotechnology and resource vision).

The proposed distinction into two traditions is used in English, where "Bioeconomy" refers to the sector and resource approach, whereas "bioeconomics" refers to the political-economic discipline.

What distinguishes both, what separates both from each other? In one way or another, both approaches to bioeconomy have a strong focus on innovations, as well as the efficiency increases of the material transformation of solar energy—the first certainly more clearly and explicitly. This includes the (biotechnological) methods of plant breeding, white and red genetic engineering, but also the bioreactors that convert cellulose into energy or basic chemical substances. The second approach focuses more on the state as a planning actor advised by science. The following subchapters illuminate the traditions and their histories of origins.

The American "The Schmidt Family Foundation" of Google founder Eric Schmidt published a future study on the development opportunities of the American bioeconomy in 2022.[1] The EU Bio-Monitor, a European project quantifying the industry, quantifies the corresponding sectors in Euro and Cent.[2] The most important databases on issues of global land use, agriculture and forestry are those of the United Nations Food and Agriculture Organisation (FAO).[3] Global land consumption and ecological footprints can be found in the databases of the Global Footprint Networks.[4]

3.3.1 An Example: The Material Flows of the Bioeconomy

If one understands bioeconomy in the sense of the understanding of the late 2000s as resource replacement, then the available quantities are a crucial piece of information. Not only the chapters on resources for construction and on urban mining, but also the following table illustrates the dimensions—and that waste upcycling will also play a decisive role. Insulation materials based on the primary raw materials sheep's wool and cotton, however, can hardly replace polystyrene in significant quantities (Table 3.2).

With regard to the construction industry, however, it should be noted that especially the agricultural "dry mass" is a largely unused raw material source. Up to now, wood has been the main raw material used. In the case of Germany, of about 48 million tons of annual wood raw material with residues and waste, around 34 million tons are used as a material (Bringezu et al., 2020, p. 18). This mass already significantly outweighs the proportion of energy-used wood (25 million tons). The total annual dry mass available from the agricultural sector is much larger than that of wood, almost three times as high (137 million tons of dry mass). It should be noted that large parts of these yields are explained by the supply of mineral fertiliser, which is mostly obtained from natural gas as a basic material and as an energy carrier through the Haber–Bosch process. Initial pilot plants for nitrogen fertiliser production using solar energy are in operation (Wang et al., 2018).

The agricultural sector traditionally sees itself as a food sector. It mainly feeds fattening animals for meat production: The majority of the agricultural dry mass is used for animal feed (around 89 million tons). In Germany, for example, almost three times as much dry mass is fed to livestock as wood is used in total as a

Table 3.2 The table compares weight units, not energy density of the materials, and is limited in its informative value to a rough comparison of fossil, mineral and biogenic raw materials on a global scale. Rounded to the nearest hundred values, except for values below 100 (According to Türk, 2014, p. 38.)

Raw material	Annual global production quantity in millions of metric tons
Waste	18,200
Coal	6,300
Crude oil	4,100
Cement	3,000
Wood	3,000
Grain	2,500
Cotton	26
Sheep's wool	2

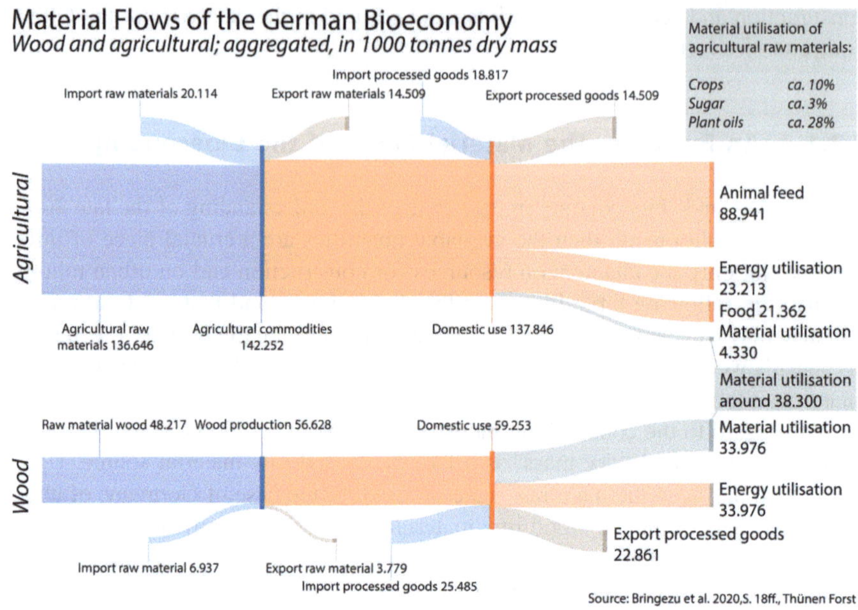

Fig. 3.4 Graphic 6. Material flows of the German bioeconomy. JG

building and material. Energetic and direct human food use (without the "detour through the stomach of the livestock") together make up around 45 million tons in the agricultural usage statistics. The material use of agricultural dry mass is still negligible and amounts to around 4 million tons, but about a twentieth of the amount of animal feed (Fig. 3.4).

Measured against the climate protection goals and the development scenarios of the bioeconomy in Germany derived from them, drastic changes in demand in the area of material and energetic use are to be expected. In a "conservative" forecast scenario of the increase, the demand for materially used "wood-like products" would increase by 0.5% annually in Germany by 2050, in the EU even by 1.7%. The demand for biomass by the chemical industry would even increase by 5.1% annually in Germany (Poganietz et al., 2020, p. 192).

3.3.2 Official Definitions and Sector Boundaries

In 2018, the European Commission presented an "Action Plan Bioeconomy" (Kershaw et al., 2021). In it, the bioeconomy is described as a path of "green growth"; the presentation appears like an archetype of the biotechnology vision and is strongly related to economic opportunities and relevant sectors, for example by pointing out the approximately 18 million employees in the sectors of the EU-wide bioeconomy, their share of the EU's gross domestic product of 4.2%. By simultaneously claiming that the European bioeconomy could "preserve nature and restore healthy ecosystems", it would "generate additional income for farmers, foresters and fishermen", restore the carbon stores of soils, forests and oceans, it almost gives the impression that it is a magic formula science capable of solving the concern of sustainability on all sides. The following passage from the summary of the Bioeconomy Action Plan, which refers to the "heart" of the bioeconomy, which is sustainability (EU Commission, 2018, p. 4), sounds almost pathos:

> To be successful, the European bioeconomy needs to have sustainability at its heart and be circular by definition. The purpose of the updated European Bioeconomy Strategy is therefore to further develop a bioeconomy that valorises and preserves ecosystems and biological resources, drives the renewal of our industries and the modernisation of our primary production systems through bio-based innovation, involves local stakeholders, protects the environment and enhances biodiversity.

This is clearly political prose. But what exactly is a viable definition of what bioeconomy is for the European legislator? According to the German definition of the late 2000s, the concept is primarily about the *sectors* that deal with renewable raw materials (Chap. 4). However, in light of the climate policy ambitions of the 2020s, it seems questionable whether this definition is still current. Shouldn't the focus be on transforming the entire economy into a sustainable bioeconomy? The EU's definition of bioeconomy from 2018—the year before the announcement of the "Green Deal"—is strongly sector-based. This is also evident in this passage from the 2018 Commission paper:

> The bioeconomy covers all sectors and systems that rely on biological resources – animals, plants, micro-organisms and derived biomass, including organic waste – as well as their functions and principles. It includes and interlinks: land and marine ecosystems and the services they provide; all primary production sectors that use and produce biological resources (agriculture, forestry, fisheries and aquaculture); and all economic and industrial sectors that use biological resources and processes to produce food, feed, bio-based products, energy and services. (Ibid.)

The sectoral approach also remains dominant in the **United States** (Frisvold et al., 2021). This approach is likely more politically consensual as it does not fundamentally "attack" energy-intensive industries, and this path also aligns with strong economic interests in agriculture and forestry. Dozens of other countries around the world have national bioeconomy strategies. Many focus on their respective resource priorities, such as forests in Finland or corn and soy in Brazil. However, in the more than 40 national bioeconomy strategies worldwide, it becomes clear that definitions are less important than the precise recording of biogenic material flows and growth contributions of the bioeconomic sectors (Thrän & Moesenfechtel, 2020, p. 41). The term bio-based economy is often used synonymously with bioeconomy. Related aspects include the approaches to a "biologisation" of technology (i.e., design based on the model of nature, bionics; Fig. 2.5) and biological transformation. This refers to the integration of natural principles into all economic sectors. To quantify the size of the sectors and their growth, the EU initiated the "EU Bio-Monitor" project (Haarich et al., 2017), but this page contained hardly any relevant content at the beginning of 2023 (Biomonitor, 2023).

The **People's Republic of China** aims to become the global leading innovation site for the bioeconomy according to the five-year plan of the Communist Party. It also defines bioeconomy sectorally, but linked with environmental protection goals, as a "model focusing on protecting and using biological resources and deeply integrating medicine, healthcare, agriculture, forestry, energy, environmental protection, materials and other sectors" (PR China, 2022). Its state bioeconomic strategy includes the construction sector. Biomass, as a crucial renewable energy source, is commonly used in various areas such as agricultural fertilizers, animal feeds, power generation for heat and natural gas supplies, and synthetic biomaterials for construction. Additionally, rice as a biomass source and local industrial park clusters play significant roles (Zhang et al., 2024).

The European Union makes a sectoral distinction between a core, partial and indirect bioeconomy ("Bioeconomy development in EU regions", 2017; cf. Thrän & Moesenfelchtel, 2020, p. 24f.):

- The **core bioeconomy**, i.e. the "fully bio-based" sectors, includes: agriculture and forestry, fisheries, aquaculture, bioenergy and biofuels, food, beverages, feed, bio-based products and processes
- The **partial bioeconomy** thus includes the "partially bio-based" sectors, which also includes construction: chemistry, plastics, construction, paper, pulp, waste management, biotechnology
- Finally, the EU understands the **Indirect Bioeconomy** "partially related to bio-based products" economic sectors, that is: mechanical engineering, technologies, services, water supply and wastewater treatment, energy, retail.
- The "resource" or "biotechnology vision"—associated with a sector focus—has a tradition especially in administrative and policy areas of research and development, as well as agricultural policy. The German Bioeconomy Council, founded in 2009, also followed this approach during its initial working period. Instead of a "grand political design," like the EU Commission's 2019 and 2025 cancelled "Green Deal" in response to the climate crisis and climate protests, the Bioeconomy Council issued policy recommendations for specific sub-areas of resource provision in its early years: land use, plant research, animal breeding, biotechnology, and aquaculture (Bioökonomierat, n.d.). Initially, the Bioeconomy Council defined bioeconomy sectorally on its website, aligning with the EU's definition; the focus was still on bioeconomy as a "growth sector" with core interests in expanding the renewable raw material base for use: Food
- Food
- Feed
- Fibre, Fuel.

Fig. 3.5 What could be learned from the spines of the cactus for a building? Question of bionics. Photos Jan Grossarth

Agriculture, forestry, food processing, fishing, aquaculture, chemical, pharmaceutical, cosmetics, textile and energy production were mentioned (see Fig. 3.5).

3.3.3 Criticism of the "Resource Bioeconomy Concept"

This approach has been criticized for leading to a purely *instrumental view* of the *nature as a resource.* This danger was early described as a "wrong path" and even as "totalitarian" (Gottwald & Krätzer, 2014). The criticism of lacking cultural-anthropological references in the concept of bioeconomy is brought up by environmental ethics. This is now also happening with the prospect that memories of man's nature-relatedness could in the future allow the bioeconomy to be understood in sufficiently broad horizons, namely in a socio-cultural (Stahl, 2022):

> However, if the potential of bioeconomy is assessed primarily in terms of profits for companies and shareholders, not only the question of the intrinsic value of nature is lost sight of, but also the fundamental, teleologically oriented question of what inventions in the context of modern biotechnologies bring "good" to mankind, as the philosopher Ludwig Siep noted. Since there is certainly more to this "good" than the satisfaction of elementary material needs and economic interests, namely also social participation and self-determination as well as the enjoyment of cultural and aesthetic values, ethical debates in the bioeconomic context should take into account the aspects mentioned in this contribution. (ibid., p. 278f.)

Technocratic ideas can also "hide" under the "cloak" of the bioeconomy (Hogarth, 2015; Mukhtarov et al., 2017, p. 1018). The already enormous land consumption of European imports of biomass in the 2010s also stimulate reflection on their sustainability (O'Brien et al., 2015). Ethical discussions about the bioeconomy acknowledge conflicts of objectives and dilemmas, addressing them independently rather than simplifying them (Zichy et al., 2014). Increased clarity in international definitions has been necessitated by their impact on local development collaboration and its legal implications (Ferreira et al., 2024). Criticisms from a feminist perspective have highlighted shortcomings in addressing the care and reproductive qualities of an economy (Hackfort & Saave, 2024).

3.3.4 Georgescu's Point of View: Radical Resource Conservation as a Political Paradigm

Looking at the historical origins of the concept of bioeconomy, it is noticeable that—as hinted at in the introduction—a development line largely different from the biotechnological one goes back decades. This is essentially an economic one. In this tradition, the dependence on and finiteness of resources is addressed as an unrecognized or underestimated basis of growth and the prosperity dependent on it.

The focus is on macroeconomic material cycles. From this perspective, the claim is derived that a loss of material qualities—such as oil, coal or minerals—must be avoided or significantly slowed down. It is also about the problematisation of the energy dependence of modern economies and lifestyles. An ideal bioeconomy is not only necessarily circular, but above all frugal and its growth would be tied to the agricultural.

3.3.5 Nicholas Georgescu-Roegen and the Degradation of Matter

This bioeconomy concept was established from the 1960s to the 1990s by the Romanian-American mathematician and economist Nicholas Georgescu-Roegen—but initially not as a political paradigm, but within the framework of economic-academic discourses, which were hardly noticed by the public. In the work of Georgescu-Roegen (1906–1996), who is considered the founder of today's "ecological economy", the term *bioeceonomics* is central. It deals, also under the impression of contemporary events such as the oil crisis of 1973, primarily with the problematic resource and energy dependence of modern lifestyles and economic systems. In his main work, he investigates the dependence of economic growth on fossil fuels, which was not recognized by the academic economics of his time (Georgescu-Roegen, 1975). The mathematician, statistician and economist, who spent most of his life researching and teaching in the United States, sought to put the unsustainable energy and resource consumption of modern economies on the agenda, especially of the economic growth theory (Maneschi & Zamagni, 1997). He directed the attention of this academic world in particular to the structurally problematic waste problems and the unsustainable resource dependence in the growth-oriented, consumer-driven economies. Economic growth theories did not take into account the fact of often irreversible material waste ("degradation of matter") at all. He attributed this on the one hand to the mechanistic thinking of economics, on the other hand to its historical backgrounds—the historical exception of an almost unlimited abundance of material:

> Two factors may explain the peculiar orientation of standard economics. First, its direct ancestor - the neoclassical school - emerged at a time when mechanistic dogma was still in great favor with philosophers and natural scientists. Second, the unique mineral bonanza enjoyed by the Western world (the home of the neoclassical school) since the beginning of the past century grew even more bounteous. When one's pantry is more than amply stocked, one is apt to ignore the source of one's fortune. (Georgescu-Rogen, 1981)

Once coal is burned, oil is burned or natural stone is broken, it is irreversibly lost for further equivalent uses. In this sense, only the renewable raw materials are "renewable", but their growth in industrialized agriculture is based on machines, oil, mineral fertiliser use (nitrogen from natural gas). Therefore, a circular economy is not Roegen's concern: This would also be part of the "mechanical" thinking in his skeptical view. Anyway, he indirectly mocks such assumptions of the majority of economists: "Thus, the economic process is envisioned as a self-sustaining, circular affair, as we find it portrayed graphically in virtually all introductory manuals." (ibid.) He contrasts his deep pessimism regarding a sustainable modern lifestyle, which he illustrated with (later rightly criticized) comparisons with thermodynamics ("entropy"), with the vision of a "bioeconomy".

3.3.6 Does the Law of Entropy also Apply to Sandstones?

This radically skeptical way of thinking about bioeconomy is associated with a long tradition of transferring the physical laws of thermodynamics to the material sustainability consideration. Thus, the energy dependence of modern economic activity and life becomes the central motivation of the bioeconomy. The second law of thermodynamics, which states that heat passes from a warmer to a colder body, but never the other way around, prompts Georgescu-Roegen not only to illustrate the energy dependence of modern life in poetic images. He even applies this law to material processes. Georgescu-Roegen derives from the second law of thermodynamics a thesis of the ultimately inevitable degradation of matter and thus a natural law of the decay of matter and calls this the Fourth Law of Thermodynamics (Thompson, 1997). This makes the decay predictions lawful. It also ultimately means that complete recycling would be impossible. Material always suffers a loss of quality when it is introduced into the economic cycle (Fig. 2.6). Thus, the entire economic and growth process is limited by the degradation of matter, with Georgescu-Roegen arguing primarily using the example of the combustion of fossil raw materials, but certainly also of waste (plastic) or (agricultural) chemicals derived from them (see quote).

> Having in mind the statistical interpretation of thermodynamics, one may argue that we can certainly reassemble the pearls of a broken necklace scattered over the floor. Is not recycling such a type of operation? To see the error in extrapolating from the molar to the molecular level, let us suppose that the same pearls are first dissolved in some acid and the solution is spread over the oceans — an experiment which depicts what actually happens to one material substance after the other. Even if we had as much energy as we pleased, it will take us a fantastically long, practically infinite time, to reassemble the pearls. (Georgescu-Roegen, 1999 [1977] p.143).

However, the theory of the degradation of matter was also rightly criticized—the transfer of the second to a fourth law would be scientifically unserious (Schwartzman, 2008; Hammond & Winnett, 2009). Also, early on, thought models of overcoming the fatal energy dependence by increasing the yield of renewable energies were opposed (Weinberg, 1981). The entropy law is invoked almost 40 years after the publication of Georgescu-Roegen's writings by authors of the post-growth movement: "Entropy is the physical measure of disorder" (Kümmel et al., 2018). However, elsewhere the exact opposite of this claim is to be read: "That it is a measure of 'chaos' or 'disorder' in a system is an example of a popular and generationally transmitted effort to assign it a meaning. Unfortunately, this shortened interpretation is an impermissible simplification of the facts" (Müller-Plate, 2020, p. 250). In any case, it is an interesting and moderately enlightening metaphor, the interpretive power of which is explained particularly against the background of Georgescu-Roegen's life and time history (Amin et al., 2012) (see Fig. 3.6).

3.3.7 Historical Precursors

The problem of the precarious resource dependence of growth and its instability was not adequately recognized or taken into account in the early years of Georgescu-Roegen's work, neither in theoretical economics nor at the government level. However, in the academic or journalistic view—beyond macroeconomics—it has far-reaching and several predecessors. Regular thematisations of the finiteness of coal reserves, which had ultimately been the driver of the Industrial Revolution from the eighteenth century onwards, have been published since the beginnings of industrialisation in Great Britain (Frischknecht, 2020, pp. 3–10). The term bioeconomy was also not invented by Georgescu-Roegen. It is rather used in different senses by earlier authors. The British biologist Hermann Reinheimer (later: Harry Ryner), a native German, used it in a publication in 1913, in which he aimed to oppose the contemporary, politically instrumentalized Darwinian competition concepts with a *humane draft* of a sociological *Bios* analogy (Fig. 3.7). He did this by emphasising those aspects of the "biological" that are based on cooperation—such as animal or plant symbioses (Baranano, 2021). In this way, he wanted to

Fig. 3.6 Degradation of matter: A handful of brick rubble in a building materials yard in Southern Germany. Photo by Jan Grossarth

counter the social Darwinism, which found scientific and political dissemination in the pre-fascist decades, with an alternative narrative.

Fig. 3.7 Historical book title and the decisive paper "En Route" (2007)

Another, distinct development line of bioeconomic conceptions goes back to Wilhelm Ostwald (1853–1932), whose so-called *Energism* can be compressed to the imperative: "Don't waste energy, utilise it". Ostwald's writings such as "Energetic Foundations of Cultural Science" (ger.: "Energetische Grundlagen der Kulturwissenschaft") are said to have influenced the former German economic politician Hermann Scheer (SPD), who is considered the "father of the energy turnaround" in Germany (Scheer, 2012). In summary, it can be said that the term bioeconomy has been used since the turn of the last century with changing emphases, and against different historical backdrops, as a metaphor for a stable, crisis-proof economy and society. Slavcho Zagorov, a Bulgarian contemporary of Georgescu-Roegen who formulated similar critiques of standard economics, has been largely forgotten in academia and the history of energy (Nenovska et al., 2024). Among the relevant authors, Nicholas-Georgescu Roegen plays the role of the uncompromising doom-monger, who speaks in frighteningly enlightening parables.

3.3.8 The Agrarian Economy as a Distant Mirror

Georgescu-Roegen's models are also based on travel experiences. Traditional agrarian economies are his target. He undertook research trips to developing countries such as Brazil or Southeast Asia in the 1960s (Georgescu-Roegen, 1960). In addition, he often refers to his observations of the peasant economy in his home country Romania in the 1940s, where he also worked as a high-ranking government official for the monarchist peasant party during the Second World War, before he had to re-emigrate to the United States from the communist regime, where he had already researched at Harvard before the war. The conclusions he drew from his travel observations led him on the one hand to a critique of the academic orientation of development economists, who focused on the development factors of economic growth, i.e. the enlargement of the gross domestic product. He himself devoted his interest to the resource and energy dependence of the developing countries—the agrarian economies as "reality without a theory" (ibid., p. 4). There he saw that growth was only possible with the institutions of the industrialized countries—banks, central banks, courts, Patent offices, large power plants, factories. According to Georgescu-Roegen, agricultural production differed primarily from industrial production in that the former was not scalable. Growth was more time-bound and the energetic inputs (sun, water, animal manure) could not simply be substituted by other energy sources, as in the case of electrically operated factories (ibid., p. 5f.). He accused Karl Marx of having made his proposals for forced agricultural industrialisation without the basis of a study of the realities of peasant economies: "Probably the greatest error of Marx was his failure to recognize the simple fact that agriculture and industry obey different laws; as a result, he proclaimed that the law of concentration applies equally well to industry and agriculture. To repeat, Marx had no opportunity to observe a peasant economy" (ibid.).

Georgescu-Roegen appears here as a mathematical economist, whose theory formation and theory criticism is based on experiences as an observer, as a traveler. From there, he attacks both Marxist and capitalist economics. His last trip to Brazil was in 1978.

3.3.9 Georgescu-Roegen in Historical Context

Georgescu-Roegen's work falls historically into the decades of the most intense exploitation of the natural environment. Since then—especially since the Rio Conference of 1992—the climate crisis has intensified, but the exploitation has slowed down, weakened, if not reversed, in some respects. Since then, states have designated many new protected areas. The increases in species extinction, biomass consumption, and other environmental indicators were stronger in the decades from 1960 to 1990 than in the following three decades, but they continue nonetheless. A partial decoupling of economic growth and energy consumption has been achieved. Although the extraction is increasing globally, and that of building minerals even disproportionately strongly. However, resource productivity has increased. In Germany, for example, raw material productivity increased by around 72% between 1994 and 2019 (Umweltbundesamt, 2022).

The oil price crisis of 1973 falls into the time of Georgescu-Roegen's work as an energy-historical shock. Also, the Vietnam War and the impressions of the Second World War, the technocratically planned and executed mass murders of the Nazi regime, and the atomic bombings of Hiroshima and Nagasaki motivated a series of influential technology- and progress-critical writings (Carson, 1962; Jonas, 1979). For example, Rachel Carson's "Silent Spring" reached a mass audience and is considered the initiator of the international environmental movement (Radkau, 2011), as it condensed the ecological damage effects of the insecticide DDT into the dystopia of the great extinction. Georgescu-Roegen's rhetoric of the "civilisational cul-de-sac" is to be read in the spirit of this time:

> A highly mechanized and heavily fertilised cultivation does allow a very large population, Pi, to survive, but the price is an increase of the per capita depletion of terrestrial resources, si, which, ceteris paribus, means a proportionally greater reduction of the future amount of life [... and...] many species associated with old-fashioned, organic agriculture may gradually disappear, a result which may drive mankind into an ecological cul-de-sac from which there would be no return.

The deep skepticism has a reason also in the biographical experience. Georgescu-Roegen experienced an early loss of his parents and the loss of his Romanian homeland, two world war and the Communist regime. After a few flourishing years at Harvard, he worked in the self-chosen "exile" of an American provincial university. His pessimism about humanity's ability to cooperate and solve problems is heavy and combative, like disappointed, in the face of naive American promises of the future—but in principle, for Georgescu-Roegen, the path of history is not predetermined to doom, but the future lies in the shaping by people: "History, as it seems to me, is a sequence of absolute surprises all explained ex post, however."

Funds-and-Flows-Model

Central to Georgescu-Roegen's work is the "Funds-and-Flows-Model", which clarifies his core concern—namely to model to what extent economic growth is materially driven (Couix, 2020). It illustrates the societal metabolism by quantifying flow and stock sizes and relating the temporal changes to each other. For example, if coal stocks flow into production (flows) and increase economic growth, the coal stocks (funds) and thus future growth bases shrink as a result. The model is part of his production theory. This also takes into account the resource dependence of growth—unlike standard production functions, which consider variables such as labor, capital and technical knowledge, but not the depletion of resource capital through dynamic outflows of the stock. This is criticized by *Bioeconomics* in classical growth theories. The Funds-and-Flows-Model can be seen as a precursor of material flow analyses and their flow diagrams, and also as a precursor of Life Cycle Assessment. But the Funds-and-Flow-Model is unlike these, a part of a macroeconomic dynamic modeling.

3.4 Tensions in Bioeconomic Construction

3.4.1 Innovation Versus Tradition

An example of the "rediscovery of the old" is the great scientific research interest in the long durability of Opus Caementitium in ancient Rome—or also in the Roman cements, which in the 19th and early twentieth century had a wide distribution as plasters, but almost "extinct" in the course of the triumph of Portland cement (Fig. 3.8). Their porosity and water resistance make them an interesting material from a sustainability perspective, and for example through gypsum additives, disadvantages such as drying crack formations can be reduced (Sanytsky et al., 2021). Further examples: The "historical innovation field" of bioeconomy could also include "unburnt artificial stones" (Haberstroh, 1922, p. 83ff.), such as clay bricks, since the energy-related emissions for these are lower. For these were used in historical construction besides natural binders like clay or gypsum also mineral fillers like sand, gravel, slag, pumice used, but also organic ones from forestry, like sawdust and cork. This is suitable to improve the climate balances as well (Fig. 3.8). If the paradigm of energy saving is in focus, traditional building methods are to be rediscovered. In Italy, churches and farms were built from rubble stone, in Denmark, traditionally small houses are built due to the scarcity of natural stone (Fig. 3.8). And the low ceiling heights of simple farmhouses in the Spessart in central Germany were an adaptation to the centuries of prevailing energy scarcity. The use of straw as facade insulation to protect against cold east winds in Eastern Galicia, which was probably regularly composted and returned to the soils, would today fall into the category of cascade use (Fig. 3.8).

Fig. 3.8 To be rediscovered? Straw outer wall insulation in Galicia, a small stone house in western Denmark, the barn roof house in Lower Franconia, Germany, and a detail of a rubble stone outer wall of a church in Liguria. The detail shows the rammed earth wall of the Alnatura building in Darmstadt. Photos Jan Grossarth (4), Delcampe

The focus on innovation is crucial and connects the different traditions of the bioeconomy. For theologian Markus Vogt, a long-time member of the Bavarian Bioeconomy Council, bioeconomy could be defined as the "innovative side of sustainability" (Vogt & Frankenreiter, 2022).

Plant breeding in particular has been the subject of technical leaps in innovation for about eighty years, but especially in the present. A growing number of selection and breeding methods, such as mutagenesis (mutation breeding) or gene markers, are to be mentioned. Since the 1980s, transgenic methods have been added, and since the 2010s, novel genome editing methods such as Crispr-Cas-9 (Sects. 5.4 and 5.5). The term synthetic biology refers to the creation of new (micro)organisms in the lab. Also, accelerations of photosynthesis through biotechnological methods are the subject of biotechnological research. The significantly reduced cost of genome sequencing contribute to increased research. The actual contributions of biotechnologies to a bioeconomic transformation are currently not yet assessable. A "large part of the innovation potential is still in the dark", after all, only a "very small part of the estimated several hundred million microorganism species living on earth are classified" (Thrän & Moesenfechtel, 2020, p. 5). But it can be assumed that not only the critical thematisation, as has been intensively done in the media, politically and scientifically since the 1990s, is an important part of the ethical debate, but also an openness to innovation in connection with sustainable development (Schleissing, 2018).

Especially in view of the high development dynamics in the field of energy technology, but also of biotechnologies since about the mid-2010s, the crisis narrative of the "energetic precariousness of the industrial way of life" should be seen more as a story with an open outcome in the bioeconomic traditions. The high dynamics of research in the field of genome research can be exemplified by the drastic reduction of those costs that arise for a genome sequencing:

Cost of genome sequencing
2006 around 14,000,000 USD
2017 1,500 Usd
2023 399 Usd
(Q. sequencing.com)

For it is particularly through biotechnological innovations that there are major efficiency advances in the use and conversion of solar energy. An acceleration of photosynthesis through precise genome interventions or even artificial photosynthesis, the creation of organisms through synthetic biology, shift the boundaries of what was unthinkable a few decades ago (Fig. 3.9). Furthermore, the productivity of a solar-powered bioeconomy and the efficiency of its processes depend on advances in the engineering areas of plant breeding, thermal pressing and bonding processes, mechanical and sensory separation processes that make waste more usable, process innovations in building material or building material recycling techniques.

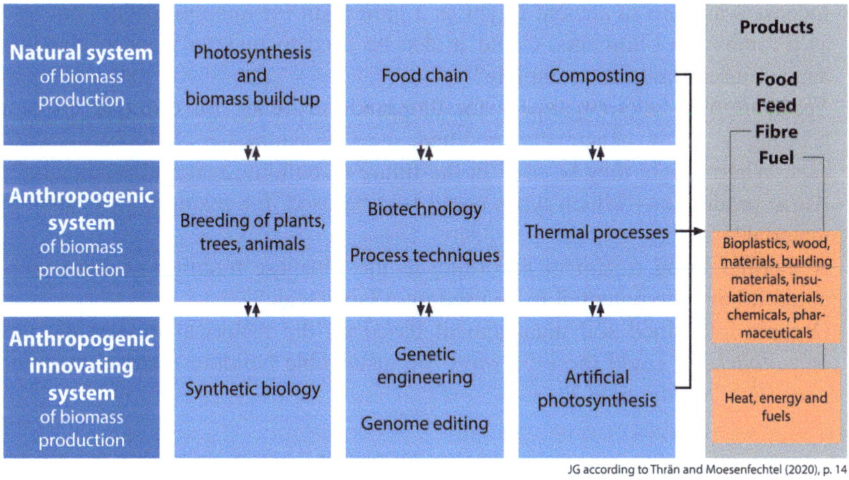

Fig. 3.9 Technological intervention depths in the human–environment relationship. JG, according to Thrän and Moesenfechtel (2020), p. 14

A "bioeconomic innovation" is often oriented towards rediscoveries and further developments of craft traditions. One reason is that these had developed under the conditions of energy scarcity (Calder, 2021). Also hints from historical building material studies on the advantageous energy balances of early established building materials can stimulate the current bioeconomy of buildings to innovations:

> At the moment, those artificial stones are of utmost importance to us, which we can use for the construction of dwellings, which are poor heat conductors and require no or as little as possible fuels for their production such as clay, alluvial, foam slag stones [...] Clay bricks [...] have been widely used since the earliest times, especially in times of great wood scarcity, in the times of need after the great wars still in the first half of the 19th century. Over time they were increasingly displaced by fired bricks, and their technique was gradually lost like that of rammed clay construction. Firmly pressed into the mold, well dried and protected from moisture, they provide cheap, dry and warm rooms with sufficient strength, especially for small houses. (Haberstroh 1922, p. 83ff., bold print JG)

For the clay bricks also known as clay patches, organic additives were also used from straw, flax or hemp shives or even heather (ibid, p. 84). Another established part of the construction circular economy has been the blast furnace slag stone for more than 100 years (ibid., p. 86f.). Xylolite or stone wood, since the mid-nineteenth century also made from sawdust, and because of its affordability also used by the Bauhaus architecture of the interwar years. That can be another field for rediscoveries. Overall, bioeconomy (of construction) can seek connections in numerous old traditions, such as these (see Chap. 4, Sect. 5.4):

- Incense was used in ancient Egypt as a gum resin for mummification, but also as a remedy in ointments. Could it also be used to make a wooden cladding against insect infestation durable?
- Yeast, used as microorganisms for thousands of years, for example, in beer brewing—is now, genetically modified, used to synthesise phytopharmaceuticals. Where might they be used in the future of building material production?
- Algae production—which types could best be used, for example, as insulation material?
- A thermal wood treatment according to the Japanese tradition of *Yakisugi*—could it have the potential to be used on a large scale?
- Organic intensified soil management based on the historical models of sods or *wölbäcker*—could these be ways for sustainable building material resource extraction "on the field"?

3.4.2 Economic Growth Versus Degrowth

Marginal Case I: Renunciation (Sufficiency, Post-Growth)

The bioeconomic orientation raises the question of whether permanent economic growth can be possible at all. It was particularly discussed academically, media-wise, and politically in the 2010s. Representatives of the post-growth economy (*degrowth*) in Germany like Niko Peach refer to the bioeconomic limits and quoted Nicholas Georgescu-Roegen. In his later writings, in which he increasingly addressed a non-academic audience, he indeed politically demanded an economy of renunciation. His rigorous "bioeconomic minimal program" (1976) included, for example, eight points:

1. Prohibition of weapons production
2. Immediate aid for underdeveloped countries
3. Gradual decrease of the population to a level that could only be maintained by ecological agriculture
4. Avoidance and strict regulation of "wasteful" energy consumption,
5. No more "extravagant gimmicks"
6. No more fashions
7. Durable and repairable goods
8. Work-life balance instead of workaholism.

While some of these points are utopian (pacifism), others are strictly dirigiste (population control within—as one would say today—*planetary boundaries*), others are illusory (no more fashions), yet others are already common ideas or political agenda in western societies (work-life balance, durable goods, immediate aid for poor countries). Some developments are emerging in the desired direction, such as the reductions in fertility rates to values of about 2 children or less on all continents except Africa (Fig. 3.10). The population growth rates have, however, strongly decreased in large parts of the world in the 2010s. A decrease in

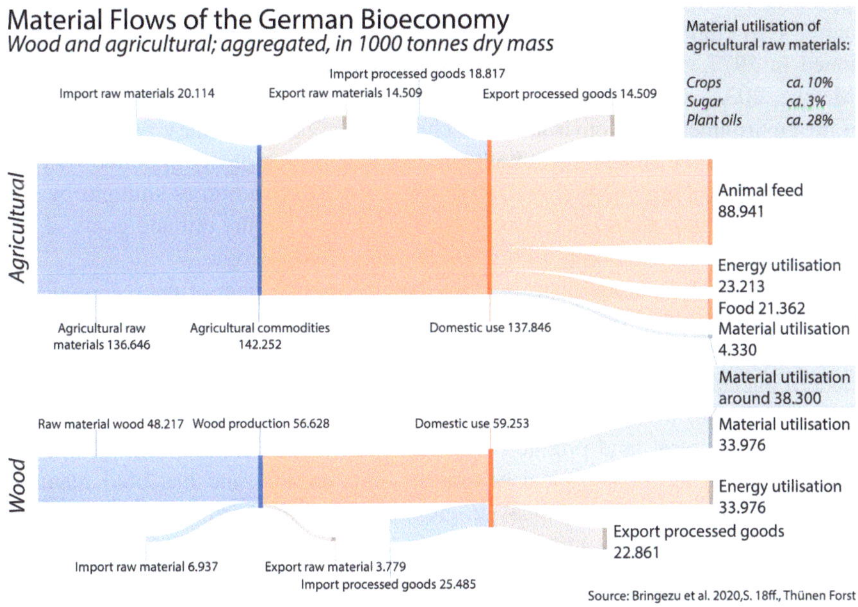

Fig. 3.10 Declining birth rates on all continents. Annika Kühn, UN 2022, p. 14

the world population by the end of the century is expected even without political population control (with the exception of China, which made its contribution) and the disasters of unpredictable world wars.

The connection of environmental crises—species extinction, soil erosion, climate change, loss of land and resources—with the prosperity and Humanity's consumption growth has long since become common knowledge. Nevertheless, per capita fashion consumption or the performance of automobiles continues to rise to the present day (Grunwald, 2017). One strategy to escape this dilemma is that of consumption reduction (sufficiency). Instead of quantitative, qualitative indicators of prosperity are proposed, such as a happiness indicator. Not only academic economists, but also a commission of inquiry of the German Bundestag, worked on this at the beginning of the 2010s, the results of which were not further pursued politically. The sufficiency approach is more an interesting conceptual idealisation than a realistically majority-capable political option—especially in times of economic crisis. It is laid out in the (late) writings of the founding father of political-economic bioeconomy.

Marginal Case II: Accelerated Innovation (Ecomodernism, Geo-engineering)

Contrary are the concepts that see the necessary reaction to the "planetary crisis" precisely in a greater technical intervention depth of humans. After all, wouldn't a large-scale technical response to a crisis caused by humans and technology be appropriate in the age of the so-called Anthropocene (Cruzen, 2002)—the "age of man"? Geo-engineering is such a conceptual orientation. The term is first mentioned in 1977, but in the context of climate policy it experiences a renaissance (Meiske, 2022). Geo-engineering is defined as "targeted large-scale manipulation of the environment. The environmental change must be the primary goal, not just a side effect, and the change must be large, continental or globally effective" (Keith, 2000). Examples include manipulations of rainfall, a reduction of sunlight by sun reflectors shot into space—or also the project to achieve the climate goals of the Paris Agreement by building with wood for two billion people.

The so-called ecomodernism is more focused on the topic of energy supply. It does not assume that renouncing consumption brings a "cure" for the planetary crises, but propagates an acceleration of technological progress. In this context, modern nuclear power plants, which one day should operate without a meltdown, play a key role, as do biotechnologies such as genome editing, which further increase agricultural land productivity—as in past centuries. This line of thought can be seen as an ideal–typical contrast to post-growth, and has bold features, whereby with maximum technical progress a far-reaching armament of lifestyles would be possible.

Excursion: The Biotechnologically Revolutionized Agriculture Without Animals—An Eco-modernist Perspective

Early human history or prehistory is about millions of years of mammoth meat, cave fires, mushrooms and berries. The fact that man went from being a hunter and gatherer to a farmer and thus sedentary, on the other hand, was usually credited to him by cultural history. The invention of agriculture was interpreted as a civilisational progress like no other. Ecomodernism changes this interpretation in the 2020s and says it's time to technologically overcome farming.

The Neolithic Revolution, however, was associated with societal progress in economic and social history. For example, with the beginning of individualization—namely the change from living together in tribes to the smaller core family. It enabled the emergence of cities and empires, writing, accounting, science, property rights, but also slavery. And indeed, economic history still does not know exactly why agriculture began about 10,000 years ago (Weisdorf, 2005). The old thesis of the great Population pressure and a dwindling of game animal populations, such as mammoths, has found new additions: Did climate changes—cold and drought—demand a settling down, as the grasses and fruits to be found in the wild became less? Or were they rather random yield increases of wild plants? Or were there prestige reasons for tribal leaders to settle down and show off with houses and fields? Was the human brain only now so far developed that it was capable of the abstract achievement of plant breeding?

Agriculture is fundamentally declared to be a problem—from the perspective of climate science. For at least one hundred and fifty years prior, the historical backdrop of the rural served as a model, while the radically technologized subjugation of nature was seen as a problem: First the growth of cities, then the material battles with nitrogen fertiliser; agrochemicals, diesel-guzzling harvesting machines, and monocultures. The farm offered an opportunity for idealisation, as it was diverse, oriented towards the preservation of material cycles, beautiful, and somehow—in the time of alternative cultures of the 1980s—also combative.

The eco-modernist self-proclaimed avant-garde of the environmental movement is now seeking to move forward. At least one of its prominent authors, George Monbiot, British-born in 1963. His "Pauline conversion" apparently took place in Finland: There he saw a pilot factory where genetically modified bacteria consume CO2 using renewable hydrogen. This is how all of humanity should feed itself in the future, Monbiot thought. "It sounds like a miracle," he reported in the *Guardian* "but it's not technically very difficult—I saw scientists in an industrial area near Helsinki turning water into food."

Now let the world be satiated by it. Pancakes, completely without eggs and milk, without cows, grass, soy feed and all the harmful stuff. Sun food. "If you change the bacteria accordingly, they make us the proteins we need for lab meat, for milk and eggs. Others make lauric and fatty acids—fish substitute from the lab. Remaining carbohydrates, after removal of fats and proteins, can replace pasta and potato chips," enthuses Monbiot (2022).

The core thesis in Monbiot's book "Regenesis" is: Agriculture must go. Especially any form of animal husbandry, not just perverted mass farming. Lab food will—fortunately for the planet—soon destroy agriculture. Then pastures and fields could disappear from the landscape, and the primeval forests and moors could become carbon sinks again, and man would operate harmlessly in his industrial areas.

The idea of overcoming climate and environmental crises through radical technological progress, however, is not so new. It has been circulating for almost ten years in American eco-modernism. The progress story of agriculture has made it possible to produce more food on ever smaller areas. Extensification strategies like organic farming are a ludicrous attempt to turn back the wheel of history. Radical technological progress in land use is needed to continue to make reserves and protected areas possible. Nuclear power should also provide the energy for this, say the eco-modernists. This idea, however, can only be sensibly justified if assumptions about a no longer increasing energy dependence of the techniques are plausible. Monbiot, an environmental protection modernist with a willingness for media-friendly sharpening, stands in this tradition. But takes the next step: to scandalize agriculture itself. After all, the machine appears vastly superior to the plant. Generating green hydrogen from solar power plants is ten times more efficient than plant photosynthesis. The efficiency of land use of the Finnish protein factory is even 20,000 times higher, because the reactors grow in height, and the artificial protein can be eaten completely, a plant only partially. The laboratory technology is also water-saving and desert-suitable.

At least one prominent environmental activist is making the radical turn away from the idea that naturalness is a compass on humanity's journey to a climate-stable future. Those who look at the technical revolutions around genome editing, fermentation and energy technology will probably have no choice but to expect major changes in food production. But isn't the condemnation of agriculture also a forgetful, pitiless view of history? A particularly big problem in the "eco-modernist" view is organic farming, which increases biodiversity in place, but consumes much more land per ton of harvest than intensive farming.

For how would we actually deal with the disappearance of cow pastures, orchards, cultural landscapes? Didn't we like them? Isn't the radicalisation of technical thinking a surrender of the idea of man as a creative shaper of the environment? In its extreme form, it seems like a media-friendly provocation, due to political polarisation also in the field of bioeconomy.

As a contrast, the reading of the old Wendell Berry is also offered. He said: "Anyone who cared about the beauty of the world has certainly not missed in recent decades how once diverse farms and yards in the USA and in Europe have transformed into thoroughly overcapitalized agricultural businesses where there is no longer any place for the traditional farmer and his world. The fences, hedges and embankments along with their diverse plant and animal life, the groves, ponds and wetlands, the pastures, hay meadows, the grassy banks of still free-flowing streams and rivers—all gone."

Good, one would have to say as a technology ecologist.

Already for the late 2020s, George Monbiot expects the third agricultural industrial revolution, after the Neolithic and the Industrialisation: "We are at the peak of the greatest economic transformation [...] While over meat- or plant-based diets are being debated, new technologies will soon make these questions irrelevant." Then the age of bacteria would begin (see Sect. 4.3).

3.4.3 Utopia Versus Historical Context

Historically, debates and discourses about the sustainability of modern lifestyles or their foreseeable decline and consequent fallbacks into agricultural or pre-agricultural civilisational states are older and follow similar patterns. The pessimism of Robert Malthus of the eighteenth century can be contrasted with the progressive hope of his contemporary Johann Peter Süssmilch (Smith, 2013), as the pessimistic "bioeconomy" of the American environmental activist and biologist William Vogt—in the face of the dwindling guano reserves—with the vigorous breeding of the high-yielding hybrid wheat by the later Nobel Peace Prize winner Norman Borlaug in the first half of the twentieth century (Mann, 2018).

The agrarian historical detailed view is like an insurance against ideologisation and polarisation of ideas of the natural. Historically considered, there have been numerous "bioeconomic innovation" since the Neolithic Revolution. Agriculture and forestry are by no means only "innovative" in the explicit age of the bioeconomy. Biomass growth is based on photosynthesis. The photosynthesis-driven plant growth has ultimately been changed for centuries or thousands of years through targeted plant breeding. The history of more efficient Land use did not begin with industrial farming systems. The long view of economic and energy history recalls the long, incremental progress story of agriculture, forestry, and food economy. This includes increases in usable harvests through innovative plow systems, crop rotations, the improvement and establishment of wind and water mills, as well as increased trade and the institutions of the bourgeois economy such as private property and jurisdiction (Jones, 2003). The relevance of the bioeconomic view on economic events—cross-sectoral—becomes clear in the historical perspective. The historical view teaches that "much was already there", even in pre-industrial times, namely about

- Land use competition
- Man-made environmental disasters
- Innovative responses to anthropogenic environmental damage.

Particularly enlightening are the contributions from the disciplines of agricultural, environmental or energy history (such as Sieferle et al., 2006). The following are two examples of anthropogenic environmental disasters of pre-industrial times and resulting technical innovations: The Magdalene flood and the decline of the culture of Easter Island.

The Magdalene Flood and the Invention of Stone Bridges

The Magdalene Flood in Europe in 1342 was not only a natural disaster (Winiwarter & Bork, 2014, p. 20f.). The fact that fertile soils were washed away on a large scale by the flood—the extreme rainfall lasted several days—was a consequence of human land use changes. In favor of fields and grassland, deforestation had taken place, the forest area in German territories in 1300 had fallen to less than 15%. In the early Middle Ages, it had been almost 90%. "Thus, the extreme rainfall in July 1342 hit landscapes often with leached soils, barely protected by vegetation" (ibid.). The consequences of the flood are partly visible to this day. In some places, the soil regenerated over the centuries through reforestation and humus formation, according to Bork in the sandy regions of the *Börde*, but in the low mountain ranges, the soil was eroded down to the rock limit and cannot be recovered. Numerous bridges were swept away. The "technical" response was to reinforce bridge piers (reconstruction of the Charles Bridge in Prague) or to build them out of stone for the first time (Balduin Bridge in Koblenz; Bauch, 2019).

Easter Island and the Rescue Through Volcanic Stone Mulch

The decline of civilisation on Easter Island is often interpreted as an example of "nature's revenge" for human overuse (Fig. 3.11). But the story of devastating deforestation there can also be "read" as a story of creative human adaptation to deteriorated, self-induced environmental changes. The European discoverers of these islands found mysterious stone gardens, more than a billion stones covering the humus-rich, good soils (Bork, Winiwarter p. 46–47). It was terraced. Only the discovery of palm roots and charcoal led to the finding that the stones had been carried here by humans and were the expression of centuries of cultivation. Thus, people responded to the scarcity of fresh water, and even managed to develop an agriculture that led to humus formation. The dependence on rainwater of Easter Island (there are no rivers and lakes there) and a complete deforestation of the

Fig. 3.11 Easter Island was home to an inventive bioeconomy that found technical means for survival. Delcampe postcard

palm forests from 1200 to 1600 AD was due to population growth and the use of palm sap, and palm wood was also burned. But The use of volcanic rock as soil protection against wind and water erosion, for water retention—that was a true bioeconomic innovation. Incidentally, it probably led to long-term mineral fertilisation. The civilisation of Easter Island died because the 15 slaves returning alive from Peru brought back a virus that was deadly to everyone else. It died, but not from "ecological collapse".

3.5 Conclusion: Bioeconomy as a Metaphor

Bioeconomy is a very interesting metaphor. Talking about bioeconomy implies an invitation to think. The term contains an inner dialectic because it is composed of two that are otherwise rarely thought together: biology and economy.

Thinking bioeconomically therefore means seeing, enduring and productively using a field of tension. It runs between the high attractiveness of the numerous "green lighthouse projects" and the attractively appearing biotechnical research vision of a "green economy" on the one hand, and climate policy ambitions on the other. But there is a third point: the inertia of change. This would be bearable from the bioeconomic perspective, better understood and comprehensible.

Bioeconomy as a metaphor is suitable for providing meaning for the economy and orientation for politics. But metaphorical concepts also bring blur, and obscure aspects of reality (Lakoff & Johnson, 2008). This explains that the concept of bioeconomy is used as well as criticized by many very heterogeneous actor groups—conservative parties and greens, wood and agricultural associations, the growth-critical and anti-globalist movements (Priefer et al., 2017). On the factual level, there are indeed many critical questions to a concept like the bioeconomy: Is there an ideology of the renewable behind it? What would be "bio" about monoculture plantations? Is there even a total economisation looming? Or: Isn't oil, purely biochemically, also "bio-based"? And so on. Bioeconomic thinking also resists the noisy promises that there are smooth or even clearly plannable paths to solving the dramatic global environmental and climate problems. Because the economic focus also implies an acknowledgment of the empirical reality of the rapidly growing world population and the desire for a lifestyle in prosperity. Under the "cloak" of the bioeconomy, technocratic ideas can also "hide" (Hogarth, 2015; Mukhtarov et al., 2017, p. 1018).

Bioeconomic thinking generates tension and friction by acknowledging and communicating the technical visions of sustainability implied in the term "bio": for example, bionics as an engineering science orientation towards natural organisms, the ideal of ecosystemically and planetarily compatible material cycles of carbon, nitrogen or phosphorus—a ultimately breathtaking planning perspective—, the justified, significant cultural guiding values of "naturalness" and "grandchild justice" of the economy. On the other hand, economic empiricism repeatedly brings disillusionment. Thus, on the playing field of the reality of everyday use, i.e., the economy, there are constant setbacks on the great political journey of transformation. These include the insolvencies of promising "Green Start-Ups" (Gatto & Re, 2021), recurring renaissances of fossil energy carriers due to geopolitical dramas like the Russia-Ukraine war from 2022, due to new oil extraction techniques and suddenly occurring new price shocks, simple resource shortages. Also, the large land consumption of European biomass imports stimulates thinking about the (O'Brien et al., 2015) increasing environmental costs of the bioeconomy. A central ethical question is the competition for land use. At whose expense does an expansion of the production or harvest of "biomass" occur? Doesn't an expansion of timber construction on a global scale necessarily mean a loss of natural forest ecosystems or a plantation use of mixed forests? Is it even appropriate to have a strong expansion of agricultural-based raw materials for construction use when there is hunger and malnutrition in the world? What can be learned from the "fuel-or-food" debate of the late 2000s, in which the political fate of biofuel at the EU level ultimately decided against the use of rapeseed and corn as fuel (Herbes et al., 2014)? The ethical discussion requires sound knowledge about raw material availability, land potential, relevant technical developments, trade interdependencies, possible biogenic, fossil or mineral raw material substitutes, but also ecological knowledge and a historical perspective that enables judgement. Ethical discussion about bioeconomy is aware of goal conflicts and dilemmas and brings them up independently, instead of covering them with simplifying moral ideas or ecological norms (Zichy et al., 2014).

The talk of bioeconomy, viewed as a conceptual metaphor, can have several functions and meanings. In Georgescu-Roegen, it refers to natural limits of renewable energy and raw materials and contains an imperative for modesty, but also a resigned undertone. The biotechnologists of plant breeding use it, conversely, as a metaphor of future hope in the spirit of Norman Borlaug—as a hope semantics of a "rescue" through further technical progress. That so much is understood under bioeconomy can be irritating: namely just two *opposing* extreme positions of polarized future discourses in the arenas of science.

But the seemingly arbitrary vagueness of the concept is also its strength. With the metaphor of the bioeconomy, there are dangers that its recipients adopt one of the extreme positions—the agricultural (negative) utopia, the technocratic "business as usual"—but the ambiguity always contains the "medicine" to ward off this danger.

Knowledgeable and reflective, it is suitable for bringing various actor groups into conversation about sustainability, who would not otherwise be, but are now thanks to this concept. But the one-sided uses and distortions—the "robbery" of an ecologically grounded understanding by biotech engineers (Vivien et al., 2019)—are also a problem. Their persistence shows that the ongoing term boom of the bioeconomy also needs ethical accompaniment, for example with regard to the goal conflicts of land use.

Finally, the bioeconomy metaphor takes the logics of action of people in the economy/businesses seriously as the backgrounds to the *conditions for better success* in the real world, instead of polemically using it as a negative foil or even caricaturing "the economy" as an antagonist of sustainability. Such a metaphorical view may irritate professionals who are trained to defend the factually correct, even seem foolish. But terms decide discourses. In the case of the bioeconomy and plural societies: they open them. And even the foolish has its function when the time is in crisis *(Cor. 1,26-31)*. The bioeconomy metaphor is an invitation and call to not slacken in thinking. Because:

> In such times of urgency, when we know we have to act, but don't know how to act, thinking is needed. Slavoj Žižek (2019)

Notes

1. https://www.schmidtfutures.com/wp-content/uploads/2022/04/Bioeconomy-Task-Force-Strategy-4.14.22.pdf.
2. https://biomonitor.eu/.
3. Fao.org/faostat/en/#data.
4. Data.footprintnetwork.org.

References

Amin, T. G., et al. (2012). Arrow of time: Metaphorical constructs of entropy and the second law of thermodynamics. *Science Education, 96*(5), 818–848.

Barañano, L. et al. (2021). Contextualization of the bioeconomy concept through its links with related concepts and the challenges facing humanity. Sustainability, 13(14), 7746.

Bauch, M. (2019). Die Magdalenenflut 1342 am Schnittpunkt von Umwelt- und Infrastrukturgeschichte. *NTM Zeitschrift Für Geschichte der Wissenschaften, Technik und Medizin, 3*(27), 273–309.

Berry, W. (1974). Speech at the symposium "Agriculture for a Small Planet", July 1, 1974, transcription by Jan Grossarth, based on the film "Look & See, A portrait of Wendell Berry" (2016).

Biomonitor. (2023). Biomonitor. Retrieved March 16, 2023, from https://biomonitor.eu/

Bioökonomierat. (n.d.). Archive of publications. Rerieved December 19, 2022, from https://www.biooekonomierat.de/publikationen/archiv.php

Bonoli, A. et al. (2021), Sustainability in building and construction within the framework of circular cities and European new green deal. The contribution of concrete recycling. *Sustainability, 13*(4).

Bringezu, S. et al. (2020). Pilotbericht zum Monitoring der deutschen Bioökonomie. CESR, Universität Kassel.

Brugge, M., et al. (2016). What is the bioeconomy? A review of the literature. *Sustainability, 8*, 7.

Calder, B. (2021). *Architecture: From prehistory to climate emergency*. London.

Carson, R. (1962). *Silent spring*, Boston.

Churkina, G., et al. (2020). Buildings as a global carbon sink. *Nature Sustainability, 3*(4), 269–276.

Couix, Q. (2020). Georgescu-Roegen's flow-fund theory of production in retrospect. Ecological Economics, 176, 106749.

Cruzen, P. J. (2002). Geology of mankind: the Anthropocene. Nature, 415, 23.

Daum, T. (2019). Of bulls and bulbs: Aspirations, opinions and perceptions of rural adolescents and youth in Zambia. *Development in Practice, 29*(7), 882–897.

European Commission (2018) Directorate-General for Research and Innovation, Bioeconomy – The European way to use our natural resources – Action plan 2018, Publications Office, https://data.europa.eu/doi/10.2777/79401

Ferreira, J., et al. (2024). A lack of clarity on the bioeconomy concept might be harmful for Amazonian ecosystems and its people. *Ecological Economics, 224.*

Frie, E. (2023). *Ein Hof und elf Geschwister, Der stille Abschied vom bäuerlichen Leben*, Munich.

Frischknecht, R. (2020). *Lehrbuch der Ökobilanzierung*, Wiesbaden.

Frisvold, G. B. et al. (2021). Understanding the US bioeconomy: A new definition and landscape. Sustainability, 13, 4. 1627.

Gatto, F., & Re, I. (2021). Circular bioeconomy business models to overcome the valley of death. A systematic statistical analysis of studies and projects in emerging bio-based technologies and trends linked to the SME instrument support. *Sustainability, 13*, 4.

Georgescu-Roegen, N. (1960). Economic theory and agrarian economics. *Oxford Economic Papers, 12*(1), 1–40.

Georgescu-Roegen, N. (1975), Energy and economic myths, *Southern Economic Journal*, pp. 347–381.

Georgescu-Rogen, N. (1981). The crisis of natural resources. *Challenge, 24*(1), 50–56.

Georgescu-Roegen et al. (Eds.) (1999). Bioeconomics and sustainability: essays in honor of Nicholas Georgescu-Roegen. Edward Elgar Publishing.

Gottwald, F.-T. & Krätzer, A. (2014). *Irrweg Bioökonomie.* Suhrkamp.

Grossarth, J. (2022). Bioökonomie Friedensbrücke? In A. Mietzsch (Ed.), *Biotechnologie Kursbuch* (Vol. 35), Berlin.

Grunwald, A. (2017) Bioökonomie—Schlüssel zu unbegrenztem Wirtschafts- und Konsumwachstum? In J. Pietzsch (Ed.) *Bioökonomie für Einsteiger*, Berlin (pp. 205–211).

Haarich, S., et al. (2017). Bioeconomy development in EU regions, Mapping of EU Member States'/regions' Research and Innovation plans & Strategies for Smart Specialisation (RIS3) on Bioeconomy, Final Report.

Haberstroh, H. (1922). Die Baustoffkunde, Erster Teil: Die Hauptbaustoffe. Sammlung Göschen. Walter de Gruyter.

Hackfort, S., Saave, A. (2024) Toward a caring and (re) productive bioeconomy? A feminist analysis of socio-technical innovations and sustainability shortcomings. *Sustainability: Science, Practice and Policy, 20*(1), 2375808.

Hammond, G. P., & Winnett, A. B. (2009). The influence of thermodynamic ideas on ecological economics: An interdisciplinary critique. *Sustainability, 1*(4), 1195–1225.

Herbes, C., et al. (2014). Der gesellschaftliche Diskurs um den, Maisdeckel vor und nach der Novelle des Erneuerbare-Energien-Gesetzes (EEG) 2012. *GAIA—Ecological Perspectives for Science and Society, 23*(2), 100–108.

Hogarth, S. (2015). Neoliberal technocracy: Explaining how and why the US Food and Drug Administration has championed pharmacogenomics. *Social Science & Medicine, 131*, 255–262.

Jonas, H. (1979), *Das Prinzip Verantwortung*, Frankfurt a. M.

Jones, E. (2003). *The European miracle: Environments, economies and geopolitics in the history of Europe and Asia.*

Keith, D. W. (2000). Geoengineering the climate: History and prospect. Annual review of energy and the environment, 25(1), 245–284.

Kershaw, E. H., et al. (2021). The sustainable path to a circular bioeconomy. *Trends in Biotechnology, 39*(6), 542–545.

Kümmel, R., et al. (2018). *Energie, Entropie, Kreativität*, Berlin/Heidelberg.

Lakoff, G., & Johnson, M. (2008), *Metaphors we live by*, Chicago.

Maneschi, A., & Zamagni, S. (1997), Nicholas Georgescu-Roegen (1906–1994). *The Economic Journal*, 695–707.

Mann, C. C. (2018). *The wizard and the prophet: Two remarkable scientists and their dueling visions to shape Tomorrow's world.* New York.

Meiske, M (2021), Die Geburt des Geoengineerings. Großbauprojekte in der Frühphase des Anthropozäns. Wallstein.

Monbiot, G. (2022). *Regenesis: Feeding the world without devouring the planet*, London.

Mukhtarov, F., et al. (2017). Away from fossil-fuels and toward a bioeconomy: Knowledge versatility for public policy? *Environment and Planning c: Politics and Space, 35*(6), 1010–1028.

Müller-Plathe, F. (2020). Keine Angst vor der Entropie. *Chemie in unserer Zeit, 54*(4), 250–253.

Nenovska, N., et al. (2024). Slavcho Zagorov (1898–1970), A forgotten pioneer of energy and ecological economics. *Ecological Economics, 226*, 108349.

O'Brien, M., et al. (2015). The land footprint of the EU Bioeconomy: Monitoring tools, gaps and needs. *Land Use Policy, 47*, 235–246.

Pern, T. (2019). Interpreting Giambattista Vico for a new psychological science: Towards a semiotic imaginative approach. *Culture & Psychology, 25*(4), 470–483.

People's Republic of China. (2022). *Bioeconomy prominent on growth agenda*, Press release. Retrieved November 29, 2022, from https://english.www.gov.cn/policies/policywatch/202205/11/content_WS627b169ec6d02e533532a879.html

Pillen, K., et al. (2020). Pflanzenbasierte Bioökonomie. In D. Thrän & U. Moesenfechtel (Eds.), *Das System Bioökonomie*. Berlin, Heidelberg/Berlin (pp. 35–49).

Poganietz, W.-R., et al. (2020). Zukünfte der Bioökonomie: Eine modellbasierte Analyse möglicher Transformationspfade. In W. Konrad, et al. (Eds.), *Bioökonomie nachhaltig gestalten: Perspektiven für ein zukunftsfähiges Wirtschaften* (pp. 187–222).

Priefer, C., et al. (2017). Pathways to shape the bioeconomy. *Resources, 6*(1).

Radkau, J. (2011). *Die Ära der Ökologie: Eine Weltgeschichte*, Munich.

Sanytsky, M., et al. (2021). Suitability of modified low carbon Roman cements for architectural restoration. E3S Web of Conferences, 280. 07002.

Scheer, H. (2012). *Der energetische Imperativ*, Munich.

Schleissing, S. (2018). Bioökonomie als gesellschaftlicher Fortschritt?—Ethische Überlegungen zur Politikstrategie Bioökonomie. *Forum Wirtschaftsethik, 26*, 70–77.

Schwartzman, D. (2008). The limits to entropy: Continuing misuse of thermodynamics in environmental and Marxist theory. *Science & Society, 72*(1), 43–62.

Sieferle, R. P., et al. (2006). *Das Ende der Fläche: Zum gesellschaftlichen Stoffwechsel der Industrialisierung*Vienna.

Smith, K. (2013). *The malthusian controversy*, London.

Soeder, D. (2022). *Mitigation by geoengineering, energy futures: The story of fossil fuel, greenhouse gas, and climate change.* Cham (pp. 205–239).

Stahl, M, (2022). *Bioeconomy from the perspective of environmental ethics, bioeconomy and sustainability: Perspectives from natural and social sciences, economics and ethics,* Cham (pp. 267–280).

Thompson, H. (1997). Ignorance and ideological hegemony: A critique of neoclassical economics. *Journal of Interdisciplinary Economics, 8*(4), 291–305.

Türk, O. (2014). Stoffliche Nutzung nachwachsender Rohstoffe. Springer.

Turnbull, T. (2021). Energy, history, and the humanities: Against a new determinism. *History and Technology, 37*(2), 247–292.

Umweltbundesamt. (2022). *THG-Minderungsziele.* Retrieved April 1, 2023, from https://www.umweltbundesamt.de/daten/klima/treibhausgasminderungsziele-deutschlands#undefined

Vivien, F.-D., et al. (2019). The hijacking of the bioeconomy. *Ecological Economics, 159*, 189–197.

Vogt, M., & Frankenreiter, I. (2022). Bioeconomy: The Innovative twin of sustainability. *Sustainability, 14*, 22.

Wang, L., et al. (2018). Greening ammonia toward the solar ammonia refinery. *Joule, 2*(6), 1055–1074.

Wei, X., et al. (2022). Knowledge mapping of bioeconomy: A bibliometric analysis. Journal of Cleaner Production.

Weinberg, A. M. (1981). Die physikalischen Grenzen des Wachstums—Überwindungsmöglichkeiten durch unerschöpfliche Energiequellen. In S. Borner (Ed.), *Sachzwänge und Gestaltungsräume des Unternehmers von morgen* (pp. 48–60).

Weisdorf, J. L. (2005). From foraging to farming: Explaining the neolithic revolution. *Journal of Economic Surveys, 19*(4), 561–586.

Winiwarter, V., & Bork, H. R. (2014). *Geschichte unserer Umwelt: Sechzig Reisen durch die Zei.* Primus Verlag.

Zhang, H. et al. (2024). Modeling denitrification nitrogen losses in China's rice fields based on multiscale field-experiment constraints. *Global change biology,* 30(2), e17199.

Zichy, M., et al. (2014). *Energie aus Biomasse—Ein ethisches Diskussionsmodell*, Wiesbaden.

Žižek, Z. (2019). *Peterson debate.* Retrieved February 24, 2023, from https://zizek.uk/slavoj-zizek-jordan-peterson-debate-happiness-capitalism-vs-marxism-transcript/

4 Renewable Resources in Construction

Abstract

The use of renewable raw materials often improves the greenhouse gas balances of buildings over their life cycle compared to conventional building materials. This can be stated even if there are numerous definitional ambiguities regarding competing concepts of climate neutrality. Life cycle assessments (LCAs) can demonstrate the relative advantage of renewable building materials. In terms of resource availability and organic material flows, the increasing use of straw from various grains as a building material is promising. Agro-ecological arguments support field crops like miscanthus, hemp, and poplar. Bamboo, flax, cotton, coconut, pineapple, reed, bulrush, hops, jute, nettle, sugarcane, and wood from the kiri tree are also considered relevant as mass building materials for the construction sector. Biotechnologically advanced applications from the lignocellulose bioeconomy will impact the construction industry. In addition to solid wood, laminated and cross-laminated timber are playing an increasingly important role, expanding the raw material base to include hardwoods that have scarcely been used in construction until now. Structural modifications of wood on the micro or nano level are also becoming relevant. This chapter provides an introductory overview.

The use of renewable construction materials often improves greenhouse gas balances compared to conventional building materials during a cradle to cradle building (product) life cycle (Fig. 4.1). This can be said even though there are numerous definitional ambiguities regarding competing concepts of climate neutrality (Lützkendorf and Frischknecht, 2020) and methodological debates about dynamic LCA approaches making discounting for emissions in $t + x$ possible. In terms of life cycle assessment (LCA), the relative advantage of renewable construction materials can be demonstrated. However, it becomes more complicated when considering how timber construction for additional hundreds of millions of

J. Grossarth, *Bioeconomy of Buildings*, https://doi.org/10.1007/978-3-031-84014-2_4

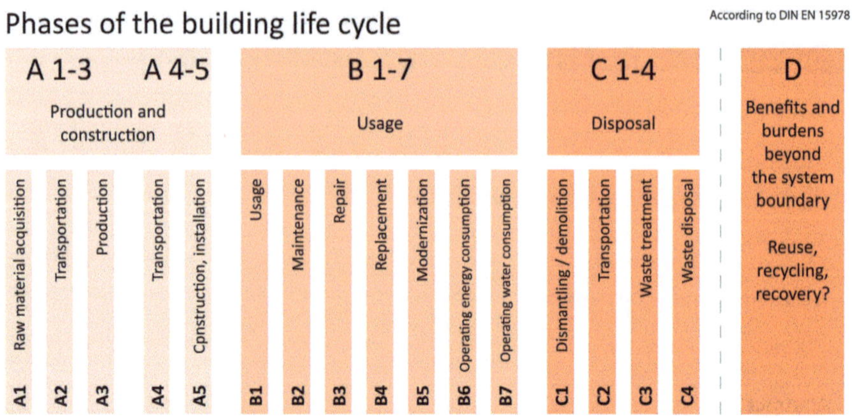

Fig. 4.1 Life cycle phases

people compared to the "status quo" would impact various environmental goals such as climate change, land use (land consumption), or water resources.

While wood is primarily used as an energy source globally (Figs. 4.2 and 4.3), its use as a material provides favorable greenhouse gas balances. There are also traditional, highly interesting application models for the engineer-timber construction—the church *Santa Maria della Saluta* in Venice, for example, rests on more than a million wooden piles (Fig. 4.2). Always covered by water, these are very durable. Tradition meets innovation. Biotechnologically progressive application examples of the lignocellulose bioeconomy will reach the construction industry. In addition to solid wood, laminated and cross-laminated timbers play an increasingly important role in timber construction. They expand the raw material base to hardwoods that have hardly been used in construction. Structural changes in the wood at the micro or nano level are added (Chap. 5). Agricultural raw materials such as grasses, cereals, and straw are also becoming increasingly relevant for construction. This chapter provides an overview.

4.1 Importance in Climate Policy

> The construction turnaround must succeed if we want to have any chance of keeping our civilisation alive.
>
> (Hans Joachim Schellnhuber (2021))

Can timber construction prevent climate catastrophe? The realisation of this urgent demand requires innovation in wood material research and in forestry and agricultural resource provision. Timber construction traditionally relies heavily on the use of solid wood, such as for structures or roof beams. Here, the strength grades (Table 4.1) are a factor for the choice of wood, the price is another. While conifers

Fig. 4.2 Built on durable wooden piles: Santa Maria della Saluta. Postcard Declampe

are cheaper and were of interest in forestry in the twentieth century, hardwoods have been and are mainly used for higher-priced furniture construction. Wood materials that have desired construction properties at reasonable prices and can preferably be used in the structural area are also a hope value of international climate policy. The CO_2 storage potentials have been calculated, while at the same

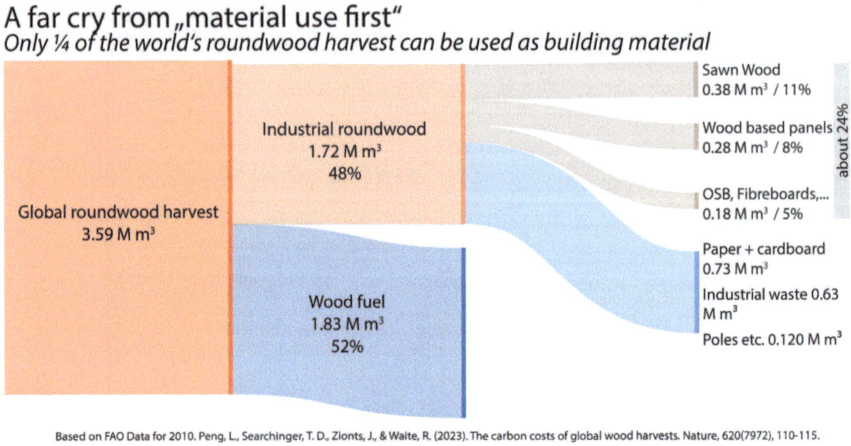

Fig. 4.3 Globally, most wood is used as an energy source

time emphasising sustainable forest management and the recycling of construction timber beyond the end of the building's life (Churkina et al., 2020; Fig. 4.4).

Renewable resources also have centuries to millennia of application histories as building materials. Reed or straw (Fig. 4.15) were used as insulation materials as well as for roof coverings, grain chaff as loose fill insulation, and straw from various types of grain was in some regions also a protective insulation against the harsh winter easterly winds: For example, pre-World War I photographs from eastern Galicia, in what is now Ukraine, show that farmhouses were protected on the facade side outside with unrendered straw (Fig. 4.5). The pictures do not reveal how the straw was attached what happened to it due to its presumably short lifespan—whether it was, for example, burned or returned to the fields as soil fertiliser. Straw and thin woods were also used in many parts of the world as reinforcement for clay plaster and as insulation against cold or heat.

Table 4.1 Compressive strengths. Q. DIN68364, categories according to (Holzmann et al., 2009, p. 90f).; Bamboo: Fachwissen Holz (2023), RWTH

Softwood		Medium and hard wood	
Poplar	32	Cherry	50
Tola	37	Oak	52
Spruce	45	Douglas fir	54
Fir	45	Birch	60
Pine	47	Beech	60
		Bamboo fiber*	> 60
		Walnut	65
		Black locust	73

* Pressed as a hardwood substitute; botanically, bamboo is a grass

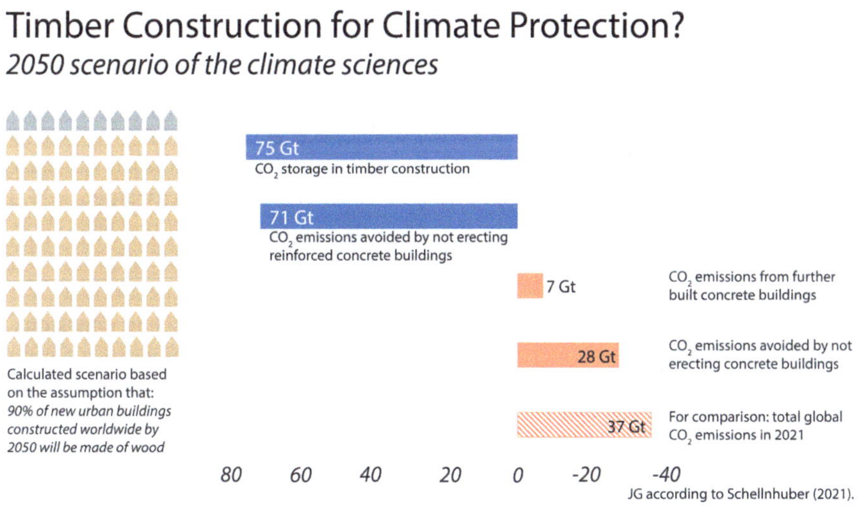

Fig. 4.4 Calculated effects of an increasing timber usage for construction

Fig. 4.5 The first straw bale house in Europe, built in 1921 in France. The resource is in the field: late summer in southern Denmark. Photo Archive, Jan Grossarth

The relative importance of such renewable agricultural building materials in industrialized countries and emerging economies only decreased as a result of the mass availability of mineral and fossil building materials such as concrete and insulation materials based on petroleum in the first two decades after the Second World War. An historically unprecedented abundance of usable energy, based on rapid increases in the extraction of oil, natural gas and coal, formed the basis for an ever cheaper usability of mineral bulk building materials (Calder, 2021). From the 1970s onwards, an end to this era was expected due to resource shortages (Bardi, 2009), but it now seems that climate protection policy is preempting this. Too with the intention of reducing the ecological, social and economic consequences of advancing climate change (Tol, 2002), the era of "cheap emissions" is politically limited. And due to their property as—at least temporary—carbon storage, there has been a greatly increased interest in application, research and innovation in timber construction since the Paris Climate Agreement in the mid-2010s. More innovation also in agricultural building materials such as straw is needed (Grossarth, 2023). With the expected increase in importance of renewable raw materials, building material traditions dating back centuries to millennia are again of interest. These include reeds, clay or grain dumps.

A look at the global distribution of land gives an impression of the size of the resource potential: More than three quarters of the habitable earth's surface are forest, pasture or savannah, thus also potential growth regions for building materials (Unnerstall, 2021, p. 68). However, it is to be expected that global temperature increases and increasing droughts can also cause forest damage in unexpected dimensions. This could reduce medium and long-term forest growth compared to historical experience (Hartmann et al., 2022). But not only the forests, pastures and savannas, but also the oceans with their algae or seagrass cultures come into question as a CO_2 storing raw material source (for example in insulation materials) for renewable building materials. Currently, these are also scientifically considered mainly in terms of their nutritional value and the question of world food (Greene and Scott-Buechler, 2022).

Renewable building materials can contribute to achieving the goals of the Paris Climate Agreement. Its core goal is to limit global warming to "well below 2 degrees" and to achieve global "net zero emissions" of greenhouse gases (GHG) by the middle of the twenty-first century. According to the German Federal Climate Protection Act (2019) and amendments to it (2021), as mentioned in the introduction, all sectors have binding greenhouse gas reduction goals to meet by 2030. Globally, climate science predicts that emission reductions will contribute the most to achieving the goals. Technical reductions such as carbon capture and storage in deep soil rock layers (CCS), or finally the reduction of emissions through land use changes (Rockström et al., 2016), contribute secondarily. In addition to many reduction approaches, the use of renewable raw materials promises that carbon will be bound in the building substance. This can succeed in various ways.

Compressive strengths (N/mm^2) of woods.

4.2 Biogenic Polymers

Biotechnologies expand the realm of possibilities for the use of biogenic building materials. Algae can not only be "harvested" in seas but can also be cultivated as microalgae in closed facilities. The use of renewable raw materials is in the context of technological leaps in development. For example, bioplastics can be produced from waste fats using genome-edited bacteria (Sect. 4.3). From a materials chemistry perspective, such technical innovations also mean, insofar as they lead to market-ready products, that compared to the twentieth century—especially its second half with its "flood" of cheap, energy-intensive produced and globally traded building materials—overall the economy will again rely more on polymeric and elastomeric resources. The proportion of metals is decreasing (Neugebauer, 2019, p. 267).

The poly- and elastomers are based on wood, agricultural and maritime biomass. In addition to traditional building materials, this can also include so-called high-performance plastics such as high-temperature polymers and bioplastics. Composite materials (hybrid materials) made from renewable and petro-based raw materials are another category. Cellulose- and lignin-containing biomasses are their most important basis. They can be used in 3D printing not only for components, but also for parts in the automotive or other industries (Mohanty et al., 2022). In this context, lignin is seen as the key molecule of a material turn towards a circular economy. The wastepaper volume as a "raw material source" is noteworthy. The global wood industry produces 50–70 million metric tons of lignin per year. Large parts of it are currently still being burned (Viera et al. 2022; see Sects. 4.6 and 4.7).

These—only cursorily outlined here—developments in the field of cellulose and lignin bioeconomy are meant when talking about the "wood-based", sometimes also the "lignin/cellulose bioeconomy" and the "agriculture-based bioeconomy".

Wood-based bioeconomy and the components of wood.

- One definition of the wood-based bioeconomy explicitly brings the raw material side into relation with a sparing, cascading use of wood: "The wood-based bioeconomy uses the raw material wood (trunk wood, crown wood and possibly stump wood, waste wood from previous use) ideally in a cascading manner. Here, the refinement steps are increasingly refined (chemical pulping and modification) and residues as well as by-products from the previous steps are used for the next stage. Only at the end the sensible use of the residues as energy carriers is then part of this process. However, the primary goal is the material use of wood, for example for building, living or for paper, and the associated CO_2 storage and substitution of fossil substances" (Miletzky et al., 2020, p. 52). The main components of wood, as well as the regrowing "biomass from the field or farm"and also from waste wood, are usable as main components in the wood-based bioeconomy:

- *Lignin*, essentially the glue between the cellulose chains; it can be used as an encrusting substance and is used as a basis for bioasphalts or bioresins.
- *Cellulose*, which accounts for the high tensile strength of the wood
- *Hemicellulose*
- *Accessory components* such as terpenes (resin acids, turpentine oil), waxes and phytosterols—for the production of resins, adhesives, wood preservatives

4.3 Case Study: Bioplastic Production by Cupriavidus Necator

The photographic evidence from the electron microscope shows *Cupriavidus necator*, an eggplant-shaped single cell (Fig. 4.6). The glutton is stuffed full of white, styrofoam-like balls, overfed like a turkey from fattening. They had fed *C. necator* fish waste (sprats). From this, the metabolism conjured up a bioplastic. Bacteria like these can do that. They metabolise waste into a valuable material. This is the science story in short form. Nice, one might say, that this is now also possible. But perhaps it is more than nice, perhaps it will be the solution. Because *C. necator* not only likes sprats, but carbon in general.

You can feed it with all kinds of waste. But can bacteria also cope with the global waste problem? Can the bacterium ensure that we can erase the word waste from our vocabulary, because every waste is then a meal? Or is this story a technology-fixated illusion that will not succeed in reality, because everything only works in the lab, but would be too expensive in implementation?

There is a lot of waste as a raw material base. In total, Germany's waste incineration plants alone swallow around 20 million tons of waste per year. This corresponds roughly to the annual plastic production of the German plastics industry. So is waste incineration on this scale a good idea? The climate balance depends on the type of waste burned: If it is wood that is burned, the effect on global warming is neutral. The combustion of plastic waste is more problematic: The plastic was made from crude oil. When it is burned, it is not much more climate-friendly than if you were to burn crude oil. Strictly speaking, it is illegal to incinerate millions of tons of waste. Because the Waste Management Act prescribes a "use hierarchy" for waste: Incineration comes in fourth and penultimate place—after avoidance, reuse and recycling. The Eco Institute says: "We still have a long way to go to a real circular economy" (Dehoust and Alwast, 2010).

But perhaps it's a short one. Perhaps it only leads to *C. necator*, and then you've already made it. Because why does humanity burn trash? Because it's disgusting. For a long time, it exceeded human imagination how mixed household waste could be "recycled". But for *C. necator*, even stinking fish aside a baby diaper is a delicacy. Household waste mixes plastic and paper with fats, savoy cabbage, cheese

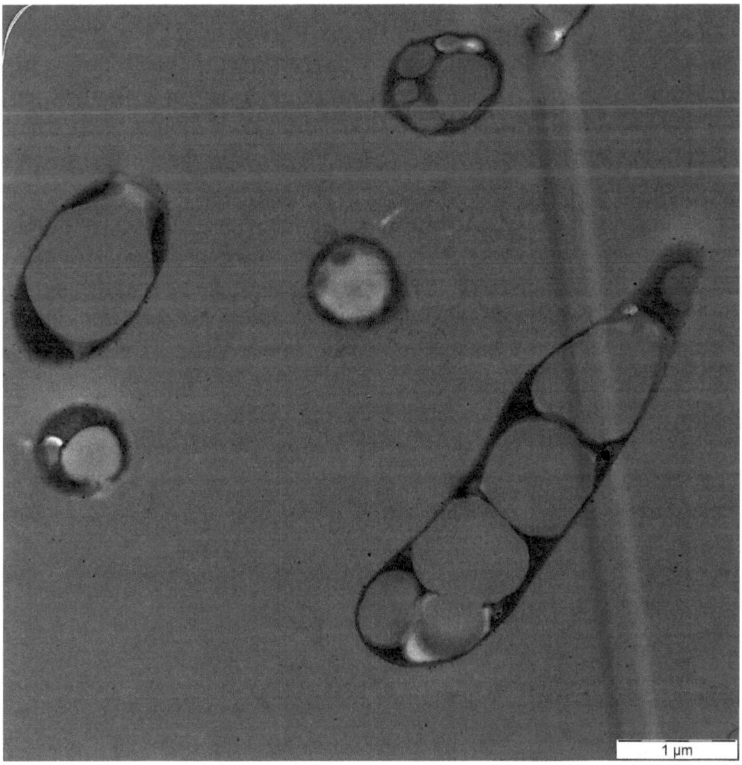

Fig. 4.6 Cuprividus necator. Photo Riedel, FU Berlin

rind, and french fry grease. Sewage sludge contains fats, feces, and medicines. Such disgusting things are not even suitable as food for insects. Therefore, only bacteria can be the saviors in need. *C. necator* is a particularly grateful guest at the dining table of civilisation. But how big is its appetite, when transferred to the big picture?

The bioscientist Sebastian Lothar Riedel knows more. He sits in his office in a brick building from the industrial founding period in central Berlin. For his employer, the TU Berlin, he leads two research projects there that deal with the new, microbial ways of waste recycling. Riedel talks about sprat heads and chicken intestines and shows chemical structural formulas—and pictures of highly interesting fat lumps that would have formed in the sewers of London. All that beautiful fat, he says. Tons of food for the little glutton. Riedel has been in league with *C. necator* for more than ten years. "You can make plastic from it, and you should," he says. PHB or PHA—polyhydroxybutyrate or polyhydroxyalkanoate in full words. Suitable for the production of bags, cutlery, and other items, these brittle, compostable materials—mixed with other plastic but for the production of almost any industrial product (Santolin et al., 2022).

How does the process work? Nitrogen, phosphorus, and carbon are needed as food for the glutton. And then comes the trick: One nutrient is given too little. When nitrogen or phosphorus is used up, C. *necator* (Latin: the killer) is in mortal fear—and metabolizes masses of carbon into biopolymers. In the end, up to 90 percent of the bacterium consists of plastic. Plastic production as panic eating. The bioplastic production with *C. necator* takes 50–70 h in the bioreactor. "I would describe *C. necator* as a super bacterium, because it can do extremely much," says Sebastian Lothar Riedel. And he can do a lot with it: "We have knocked out genes." That means, the metabolism has been genetically modified. Using so-called genome scissors like Crispr-Cas9, metabolic pathways can be specifically modified to produce PHAs with tailor-made composition and thus adjustable properties. This is the crucial step for the production to become economical. In addition to genetic modifications, the type of food also certainly determines the quality and quantity of bioplastic. Any food that contains carbon is suitable. A new study by scientists from Zhejiang University in China in the journal "Biotechnology for Biofuels" provides a global overview: Bioplastic has already been successfully made from dairy waste, used oil, food waste, lemon peels, brewery waste, or waste from paper and cellulose factories. Pure CO_2 can also be fed as a carbon source, as can microalgae (Fig. 4.7). Various bacteria are used as feeding organisms. Lipids can also be produced in this way.

Fig. 4.7 Microalgae cultivation in Wageningen, Germany. Photo by Jan Grossarth

Basic research has advanced. Now there is a wave of start-ups. TU Berlin is cooperating with Animox, a Berlin biotechnology company that manufactures plastic granules from slaughterhouse waste and *C. necator*. This then goes to *Cuba*, an injection molding company in Neuruppin in eastern germany. From this, custom-made plastic products are made for industrial customers. The bioplastic is quite expensive, but there are special demanders: companies, for example, that urgently want to reduce their CO_2 footprint. Also from areas where plastic is banned (bags, disposable dishes). However, the danger of greenwashing is always present. The smallest additions of bioplastics can be marketed as "green". The textile industry is also paying attention to the new processes. A research project of the TU Berlin with the Institute for Textile Technology of the RWTH Aachen ("PHAtex") is taking care of this.

It never works without fermenters. They are the bathtub for bacteria and nutrient substrate. Fritzmeier Environmental Technology, a company south of Munich, manufactures such fermenters. "The application potential of the fermenters is very large," says research director Michael Finke. However, he has to disappoint expectations of quick world salvation: currently, most customers are not feeding the bacteria with waste or sewage sludge, but with corn, rapeseed or sugar. "Because the cleaner and simpler a substrate is, the better," says the biotechnologist. Waste is complicated. Two years ago, the Japanese company *Kaneka* opened a microbial bioplastic factory, shortly before *Danimer Scientific* from the United States inaugurated a PHA factory in Kentucky. The two feed their bacterial strains with waste streams from palm oil and rapeseed oil factories.

The next step in the circular economy is already emerging: bioplastic is made from waste and sewage sludge. *Full Cycle* is the name of an American start up that wants to transform municipal food and other organic waste into PHA on an industrial scale in the near future. The bioplastic could be reintroduced into the organic cycle again and again. *Full Cycle* holds a patent for an industrial process for microbial waste recycling. Another pioneer is located in the Netherlands, the global paradise of the circular bioeconomy. The fresh start-up *Paques Biomaterials* wants to help customers use industrial wastewater streams to produce bioplastic. It promises on its website thanks to its own technology "a tenfold higher economic efficiency" than is currently possible in wastewater recycling.

In Germany, the *Emscher Genossenschaft*—one of the largest sewage treatment plant operators—is one step ahead of most wastewater companies. As early as 2015, it began testing the microbial use of carbon-rich waste streams together with the *Brain Biotech AG* from Zwingenberg in southern Hesse. But the result was not plastic, but lipids, i.e. fats for the chemical industry. The lubricant manufacturer *Fuchs Petrolub* made trial additives from this. Now the process is actually supposed to grow on an industrial scale. It is currently being transferred to "a recycling scale," explains Guido Meurer, who is head of the department for the development of bacterial strains at *Brain*: "The additives show astonishing properties in technical application." They are biodegradable.

Even old socks could become an interesting raw material, thinks *Brain*. They remember: *C. Necator* is not deterred by anything. *Brain* therefore wants to be ready soon. The head of the company's business development, Martin Langer, says: "Socks often contain nanosilver today, which could be extracted with the help of optimised microorganisms or enzymes." Already now, *Brain* produces an active ingredient from fermented orange peels that has an antimicrobial effect and could replace conventional preservatives in the future. Initial approval as a food ingredient has been applied for, initially in America.

The industrial demand for these substrates is crucial—as is the case with bioplastics. Whether our waste continues to be incinerated is a question of price. The cheaper the price, the greater the demand for bioplastics will be. Then companies will invest in large, industrial fermenters—and thus in the mass cultivation of *C. necator*. Currently, a kilo of PHA costs between five and twenty euros. A kilogram of plastic made from crude oil is cheaper than one euro. The chemical industry produces more than 360 million tons of plastic worldwide every year. The contribution of bacteria is ridiculously small. The world production of PHA and PHB was 36 thousand tons last year. For *Cupriavidus necator*, the feeding has just begun.

Further literature on the application fields of C. necator:

- Biofuels from C. necator based on CO_2

Panich, Justin, Bonnie Fong, and Steven W. Singer. "Metabolic engineering of Cupriavidus necator H16 for sustainable biofuels from CO_2." *Trends in biotechnology* 39.4 (2021): 412–424.

- C. necator as protein feed for edible insects

Chee, Jiun Yee, et al. "The potential application of Cupriavidus necator as polyhydroxyalkanoates producer and single cell protein: A review on scientific, cultural and religious perspectives." *Applied Food Biotechnology* 6.1 (2019): 19–34.

- Bioplastic from CO_2 using C. necator:

Stöckl, Markus, et al. "From CO_2 to Bioplastic–Coupling the Electrochemical CO_2 Reduction with a Microbial Product Generation by Drop-in Electrolysis." *ChemSusChem* 13.16 (2020): 4086–4093.

- Economic viability of using C. necator

Pavan, Felipe A., et al. "Economic analysis of polyhydroxybutyrate production by Cupriavidus necator using different routes for product recovery." *Biochemical engineering journal* 146 (2019): 97–104.

Bioplastic: Does "biogenic" always mean "degradable"? Renewable or biogenic does not mean that a (plastic) material is also biodegradable. Biodegradable polymers include biopolymers such as polylactic acid (PLA) and polyhydroxyalkanoates (PHAs) or proteins, as well as some petrochemically produced aromatic polyesters. Persistent, i.e. not quickly degradable, are the petrochemically produced bulk plastics such as polyolefins like PE, PP and PVC, PS or PU. They remain in the soil and seawater of the oceans for years to decades or centuries. But biogenic substances can also be persistent, like polymers from oils, Sheep's wool, and to some extent also lignin. Amber is an example of persistent biogenic materials. (Türk, 2014, p. 55).

4.4 Innovations in Plant Breeding

The term *genome editing* refers to a series of novel and precise genome modification techniques. These have been increasingly discovered and researched since the beginning of the 2010s and are also applied in the agricultural and horticultural sectors. Since then, the range of fruit and vegetable varieties modified by Crispr-Cas-9 and similar methods has expanded from initially tomatoes and oranges (2014) to cucumbers (2015) and melons (2016) to kiwis, bananas, cocoa, papaya and others (2018; cf. Wang 2019). In the context of the bioeconomy, such methods can also contribute to increasing biomass availability.

Genome editing techniques like Crispr-Cas9 make it faster and more precise to achieve desired plant traits than conventional breeding techniques or mutagenesis. The first plants are being cultivated outside of Europe. Disease or fungus resistance, growth rate, root length, drought resistance are some of the breeding goals (Ratner et al., 2016). Goals of resistance breeding have already been achieved.

Trees, on the other hand, have received less attention as a research object. In plantation farming, applications appear plausible in the medium term (Bewg et al., 2018), for example with fast-growing poplars, eucalyptus or bamboo. In the context of climate policy ambitions to make the construction industry a carbon sink, genome editing can make contributions. The cellulose or lignin content of trees or grasses can be changed as needed. But it is not trivial to replicate the "breeding successes" from the greenhouse (Fig. 4.8) in the field or vice versa (Chanoca et al., 2019; Jang et al., 2021; Fladung and Kersten, 2022).

Green genetic engineering (plant breeding for food purposes, but also forestry).

- Molecular breeding
- Transgenic plants (in the field from the 1990s, initially corn, then cotton, sugarcane, many vegetables, poplars)

Fig. 4.8 What's growing there? View from a greenhouse in Jordan on the border with Israel. Photo Jan Grossarth

- BT toxin
- Enrichment with micronutrients (orange maize)
- New breeding techniques, such as Crispr-Cas9
- Adaptation of the plant to its industrial use
- E.g. changing the lignin content of wood (poplar) or grain.

Red genetic engineering (medicine, veterinary).

- Cell factories (microbial factories): e.g. Taxol production in bacteria
- "Plantibodies" as an alternative to vaccines.

White genetic engineering or biotechnology (industrial biotechnology).

- Fermentation, other biocatalysis
- Technical enzymes
- Cellulase, lipase, protease, amylase, phytase, xylase,..., synthetic biology ("programming" of bacteria).

The beginnings of plant genetic engineering lie in the food sector. Initially, they produced resistant breeds of tomatoes or cereals (from the 1980s onwards). With trees—initially this was, and remains until the beginning of the 2020s, the poplar—there are similar efforts. But in the context of sustainability, the breeding goal is now a different one: Here, it is also about an increased genetic diversity of forests, and also of species and species within tree genera (see Sect. 4.6). For example, with regard to site adaptations, genome analysis, but possibly also biotechnological genome changes, are attributed possible positive contributions in addition to risks at various levels (Fernandez i Marti and Dodd, 2018). A specific technical obstacle to application is the usually extraordinarily large genome length and complexity of important commercial tree species (see Cao, 2022). Of the 700 plant species sequenced by the end of 2021, about 200 were trees and shrubs. The smallest tree genomes are those of Indian sandalwood and the peach tree. The largest are those of the economically significant spruce, the white spruce, as well as pine and fir species.

> The threats to biodiversity posed by climate change underscore the need for a better understanding of the genetic basis of phenotypic traits. In sessile, long-lived tree species, this will be of utmost importance, as the success of populations is likely to depend largely on existing genetic variation. The latest gene editing technologies (Crispr-Cas9) promise an elegant approach that will take the genomics of forest trees to the next level by enabling rigorous testing of gene function and its role in tree adaptation to their environment. (Fernandez i Marti and Dodd, 2018)

4.5 Timber Resources and Forestry

4.5.1 Global Forest Area

Carbon stocks in terrestrial biomass exceeds marine by far (Fig. 4.9). Forest are a main source. How large are the timber resources that will or can regrow? Worldwide, about one third of the land area or 4.06 billion hectares are forest—and of these 45% of it tropical and 27% boreal. Thus, in 2020 there was 0.52 ha of forest per person on earth.

- 93% of this was naturally regenerating forests, and
- 7% plantation forests.

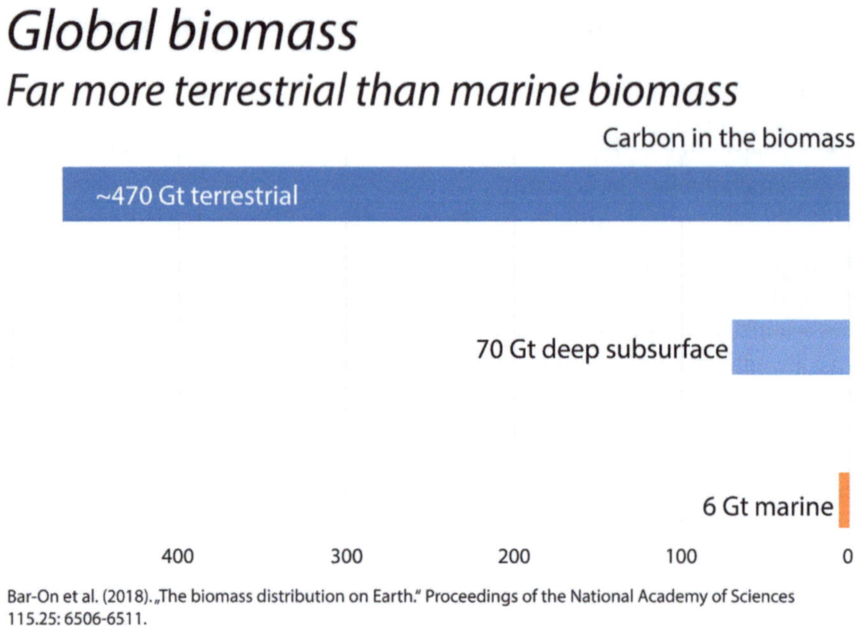

Fig. 4.9 Terrestrial biomass carbon stocks exceed marine by far

The proportion of plantation forests, however, is increasing, especially in Asia. Although forest area continues to be lost in favor of agriculture or urban and infrastructure growth, the rate of forest loss has been decreasing since 1990 (FAO, 2020). However, there is no talk of a globally uniform trend. Some parts of the world have growing, others shrinking forest areas. In Asia and Europe, forest areas have been increasing again since the 1990s. In South America and Africa, however, they are significantly declining. Globally, the resource wood is distributed to 54% on only five states:

- the Russian Federation (20% of global timber resources)
- Brazil (12%)
- Canada (9%)
- The United States (9%)
- China (5%).

4.5.2 Global Reforestation Potential

The size of reforestation potential is disputed. The so-called Crowther study by ETH Zurich estimated the global potential of degraded land that could be converted into forest areas with minimal or no land use conflicts to be around 900 million hectares (Bastin et al., 2019). This would be almost as much as the total land area of the United States (983 million hectares) or about 90 times the forest area of Germany (about 11.4 million hectares). On the one hand, according to the Crowther study, such reforestation would restore local water cycles, enable local value creation through wood use, and increase local biodiversity. On the other hand, usage conflicts with grazing by nomads are to be expected. These conflicts could not be identified by the satellite images evaluated for the study. Nomads traditionally use such areas as grazing land. And where would these "new" trees grow? An increase in forest areas would be necessary on all continents by 2050, but disproportionately again in the United States, China, India, Brazil, and Canada.

Another or complementary approach to reforestation would be to reduce the burning of wood for heating purposes. Electrification of heating would be necessary for this. But also more energy-efficient (or smaller) buildings would contribute. Especially in Asia and Africa, but also in South America, the volume of heating wood used per capita is higher than the volume of roundwood used. About half of the wood harvested globally is used as firewood or charcoal for heating. This is probably a "sin" when measured against the ideal of a circular bioeconomy.

4.5.3 Forests and Climate Change

When considering future wood resources, the expected change in the species composition of natural forests due to climate change must also be taken into account. According to forecasts, the spruce will have disappeared from European forests by 2070 at the latest, beyond the alpine highlands and Scandinavia. But the beech forests will also be retreating by then. Instead, the proportion of oak forests is likely to increase. Mediterranean oak species or Canadian Douglas firs are considered to be the European woods of the near future warmed by the earth. They would have to be planted specifically. The pace of natural tree species migration cannot keep up with the speed of global temperature changes (Pearson, 2006). The limiting factor of future forest growth is mainly the lack of precipitation, less the rising average temperature (Lippe, M. et al. 2017, p. 44).

4.5.4 EU Forestry Research

The European Union has significantly promoted the cultivation and use of wood as a raw material within the framework of the "Horizon 2020" funding program (Research and Innovation Forest). Most recently, the EU provided about 460 million euros for this (2015–2017). This was supplemented by the EU wood construction initiative "Build in Wood" (2019–2023). It is advertised with the catchy number that every cubic metre of regrown wood stores one ton of CO_2. Wood is ideal in the innovative sense as a building material for the prefabrication of large components, it contributes to waste reduction in construction, has a good ratio of weight to strength—and the European forests have been growing for years: As a result, they provide on average more than 800 million cubic metres of wood per year (EU, 2023).

4.5.5 Sustainably Certified Forestry

Sustainable forest management has been made visible and marketable for many years through established certificates such as the *Forest Stewardship Council* (FSC). More than two thirds of the Germany's forest area is certified for sustainability. The FSC seal was one of the first one of the first sustainability certificates ever. Certificates for sustainable construction require the use of such woods. In addition to ecological criteria, the FSC seal requires an improvement in working conditions and the protection of traditional forms of forest use, with each state defining its own adapted criteria based on a catalogue.

4.5.6 Tree Species Migration

The forest boundary, for example of birch, spruce and pine, is shifting in the direction of the polar north (for example in Norway, Russia, Alaska), and Mediterranean regions are desertifying. The climate effects are ambivalent; rising CO_2 concentrations in the atmosphere also act like a fertiliser, as they increase the water use efficiency of trees and allow them to tolerate drier soils (Wang et al., 2020). The forest losses in the global south can probably be "more than compensated for [...], leading to a world with more and faster growing trees in the coming decades" (Ibid). CO_2 fertilisation and biomass build-up are closely linked.

Forest productivity in Central Europe is predicted to decrease due to climate changes. Projections for Bavaria, for example, range from a 1.3 to 18% decrease in biomass yield. An increased genetic diversity of forests is considered a forest ecological patent remedy, to make these more resilient (Bartsch and Röhring, 2015, p. 159ff).

Species distribution models predict a comprehensive redistribution of tree species in the next century (Aitken et al., 2008). The extent to which populations adapt depends on phenotypic variation, strength of selection, fertility and other factors. On their "great migration north," trees due to their fertility starting only after years and very long life cycles cannot keep pace with animal species. Especially long-lived trees are at an evolutionary disadvantage. An exponentially accelerated species extinction also of wild birds, amphibians and also plants is expected in the twenty-first century compared to earlier millennia (Pereira et al., 2010; Fig. 4.10).

Fig. 4.10 Before the migration north? Oak forest in Liguria. But the natural migration of tree species cannot keep up with the pace of climate change. Photo by Jan Grossarth

4.5.7 Drought-Tolerant Tree Species

The planning horizons of forestry are long. The planning basis is complicated, as climate change makes the growth conditions of the future an equation with many unknowns. Late frost tolerances like drought tolerances will become more important decision criteria for the choice of a species. For example, late frost tolerance in beech and silver fir is particularly high, here birches and Scots pines, on the other hand, perform poorly. Another example is drought tolerance. Of the native tree species grape oaks and Scots pines or black locusts are considered particularly drought tolerant, birches or spruces on the other hand as little (Bartsch and Röhring, 2015, H p. 83f). The beech has advantages over other tree species in Central Europe. That's why it has spread so strongly over three millennia: it is relatively shade, late frost and about medium drought tolerant, its weakness however is the susceptibility to fungi due to prolonged wetness; it tolerates a wide range of soil qualities and acidity levels (Bartsch and Röhring, 2015, pp. 83–88). Nevertheless, even the robust red beech is in question in the face of extended dry periods like heavy rainfall.

In North America, the North American Species such as beech or sugar maple are widespread, as they can better cope with the seasonally more pronounced temperature fluctuations. The adaptation of foreign species is increasingly and controversially discussed in forestry circles. Unwanted harmful effects in complex ecosystem structures should be avoided. An example would be the lack of leaf-decomposing microorganisms of an invasive tree species. In Europe, Douglas fir, grand fir, and red oak are particularly being discussed as climate-adapted, non-native tree species suitable for cultivation (Ibid., p. 153).

4.6 Innovations in Timber Construction

4.6.1 Pulp Materials

Wood resources and raw and residual materials of the wood-based bioeconomy are not only applicable to timber construction. Other sectors also have an interest in them. In particular, the raw and residual material sector connections from the paper to the construction industry are likely to increase in the future. In innovation studies, the paper industry itself mentions structurally usable paper materials such as:

- Honeycomb sandwich cores for lightweight construction, whose odor absorption capacity and sound insulation are highlighted (Paper Technical Foundation (PTS) 2015)
- High strength nanofibrillated cellulose
- Facade insulation made of weather-resistant paper
- Paper PV module carriers for roofs
- Concrete reinforced with paper fibers ("paper concrete").

Building with paper has already been tested in smaller individual cases as a supporting structure—for example in a Japanese garden pavilion; larger scalable application fields are expected (Knaack, 2022). How these innovations are to be assessed from the perspective of the circular "ideal" of purity of type and separability will be seen. Further wood-based innovations are adhesives and synthetic resin made from lignin. The following chapter deepens the innovative bioeconomic, lignocellulose-based building materials.

4.6.2 Glulam and Cross-Laminated Timber

Various innovative wood materials are attributed with special performance. This is the case, for example, with CLT (Cross Laminated Timber), which is made of cross-laminated boards. They are not only usable in solid wood construction or for ceiling constructions, but also in combination with concrete as hybrid ceilings. Cross-laminated timber, on the other hand, consists of parallel glued boards. Laminated veneer lumber (LVL) are thin glued wood veneers. From the point of view of separability and purity of type, nailed stacked timber is also interesting, even if it is used less frequently than the other three materials.

CLT is referred to as a "game changer" because it makes wood building materials more attractive for multi-story buildings and engineering construction. Glued woods also make hardwood and thinner softwood suitable for the construction of structural walls, from ceiling and roof elements up to manufacturer specifications to 16 m in length, 3.45 m in width, 35 cm thickness with strength standard C24, using formaldehyde-free adhesives (Storaenso, 2021).

The allegedly world's first hardwood CLT building, which was opened in 2017, was the *Maggi's Cancer Care Centre* in Oldham, UK. For the cross-laminated timber, the wood from the American tulip tree, a type of magnolia. The material produced from it has a better ratio of tensile strength to material weight than concrete and steel (Americanhardwood, 2023). Later, also in Great Britain, the *Kingsdale School Sports and Music Hall* was built. An early CLT roof construction was received by a conservatory in Sheffield, UK, as early as 2007 (PRS, 2023). A "later" CLT structure that received attention in the context of an outstanding overall architectural achievement is the museum *Andersen Hus* built in 2020 in Odense, Denmark (Fig. 4.11).

High-rise buildings have also been constructed worldwide from CLT for several years. Examples include *Mjostarnet* (2019) in Norway, which is 85.4 m high (Dach & Holzbau, 2019), or the *Woho* residential tower in Berlin Kreuzberg, a wood-hybrid construction with a planned height of 100 m (UBM, 2022).

Fig. 4.11 The beauty of laminated timber: Foyer of the Andersen Hus in Odense. Photo by Jan Grossarth

Compared to solid wood, CLT offers promising usage options for load-bearing components of multi-part structures. But above all, it expands the raw material base. Since the beginning of the 2010s, birch and beech have been the focus of corresponding research on CLT applications (Fig. 4.12). Not only in the *Maggi's Cancer Care Centre* was tulip tree used, which is called Yellow Poplar in American (Espinoza and Buehlmann, 2018). Only a few research works have investigated the SRC-suitable poplar, especially the "Robusta" was researched here—sometimes cross-glued with other species such as walnut or chestnut, which would allow the use of such woods from agroforestry systems to be valued (ibid.; Sect. 4.10).

While the production of CLT is still predominantly concentrated in Central Europe and Scandinavia, where it was also developed, it is assumed that the majority of growth from the 2020s will take place outside this continent. CLT is still predominantly made from spruce, pine, larch and Douglas fir, but there is increasing interest in the use of hardwood species (Espinoza and Buehlmann, 2018). For example, the manufacturer *Pollmeier* in Eisenach, Germany, has been producing beech veneer laminated timber for several years and markets it as "BauBuche" (Sect. 4.9) which is roughly translating to "ConstructionBeech". *Hasslacher Norica*, an Austrian forest product company, has built a house out of birch CLT in Lemberg, Austria. And *Waugh Thistleton*, an architecture firm in the UK, had a factory and the headquarters of the furniture office *Vitsoe* built from beech CLT.

Fig. 4.12 Laminated timber from the poplar: Suitable for higher tasks? Photo by the Institute for Timber Construction at the Biberach University of Applied Sciences

There are also ecological and long-term economic considerations that speak for increased use of hardwood. This is about the changed composition of some forests in Central Europe, where beech and then oak species are expected to increase. Such CLT is also seen by climate science as a "game changer" for timber construction in competition with concrete. In Asia, CLT with bamboo components is particularly expected to have a great future. Botanically, bamboo is actually a grass and grows much faster than forest trees.

4.6.3 Case Study: Beech as Construction Wood in Thuringia (Germany)

The forest is full of history. This can be seen, for example, in the entrance hall of the large sawmill in Creuzburg, Thuringia. Here are the finds from past years on display—taken from tree trunks that had to be sorted out because old metal parts were stuck in the wood. Among the trees, sawmill employees still find, for example, grown-in horseshoes, cartridges, bomb fragments or insulating parts of power cables. Beeches can become hundreds of years old; the finds bear witness to early industrial times, war crimes, reconstruction. But the trees themselves also tell stories of the zeitgeist. In the German Federal Forest Inventory (ger.: Bundeswald-inventur), the major report of the federal government, which appears every ten years and was published a few weeks ago in its third edition, there are age pyramids and population diagrams. What they reveal: The German forest is getting older overall. And "greener," so to speak. The young forests are in the latest forest inventory already classified as very close to nature or close to nature by 51%, the mature forests ("main stock") only by 36% (data from 2014).

Close to nature, that means: There is more old wood for the birds, the trees are getting older and thicker, and there is especially more and more deciduous and mixed forest and proportionally less coniferous forest. This is what the increasingly ecologically trained foresters and forest politicians want. They want more animal species to live in the forest, the soil quality to increase, the forest to become less sensitive to orcane damage or bark beetle plagues. The mixed forest provides all this. The change in forestry policy has been particularly drastic in the area of the former German Democratic Republic (GDR): while forestry students there were still learning the techniques of clear-cutting and rapid reforestation, today natural regeneration, careful densification and ecological balance are on the curriculum throughout the country.

So at some point in the future, there will be more deciduous than coniferous tree area for the first time in decades. This was already the case in earlier times. About 100 years ago, the ratio then tipped in favor of the conifers. Because they grow faster and can be better processed industrially. The fact that there are so many pines and spruces today is mainly due to the post-war period. Because most trees that are felled today are children of the fifties and sixties: At that time, it was consensus that the economy should grow. So people also planted in the forest what grew quickly: spruces in the West, mainly pines in the GDR. 25% of all trees in the German forest are today spruces, 22% pines. Sawmills and the wood industry can process these trees earlier, and with pines, large open areas, like those created during the war, can be reforested more easily. A beech, according to the latest forest inventory, increases on average by about 10 cubic metres per hectare per year, a spruce by about 15 cubic metres.

Since the eighties, ecological thinking has also found its way into forestry. Increasingly more deciduous trees were planted. Among the one- to 20-year-old trees, more deciduous wood has now been mapped in the German forest for the first time again. This affects the wood industry: Because it has adjusted over the decades to more spruces and pines regrowing. The sawmills are set up for this. Most are only made for the—softer—softwood. "Beech grows slower, the performance of the German forest is decreasing," says Lars Schmidt, Managing Director of the Sawmill and Timber Industry Association.

While some complain, the first major company in the industry has decided to adapt to the trend towards ecological reforestation. It is *Pollmeier* in Creuzburg. It is the first to manufacture a new building material from deciduous wood, from which halls, wind turbines or skyscrapers can be built. Here in Thuringia, near Eisenach on the former inner-German border, there is now the at the time of its foundation the first sawmill in Europe, which can process beech wood into building materials on a large scale. The thick beams, which are up to 35 m long and are exported to the world, are called veneer laminated timber. *Pollmeier* invested 105 million euros in the plant, which opened in the summer. Here, the trees are first boiled, then peeled from the outside to the core. The thin plates that are created in this way are dried, glued, and massive beams are created that are so solid that you can allegedly also build skyscrapers with them. Otherwise, hardwood was mainly taken up industrially by the furniture, pulp or parquet industry, if at all. However, the Germans burn a large part of it in their fireplaces. Hardwood was not suitable for construction. Even here in western Thuringia, the forests are full of beech trees. And beech wood should, according to the law of supply and demand, become even cheaper in the future. According to the forest inventory, no type of tree grows as fast as the beech. On the other hand, there is less and less spruce wood.

4.6.4 Fast-Growing Poplar for Construction

Poplar wood from short rotation plantations (SRP) or agroforestry systems are either converted into electrical power in biomass or heating power plants, or materially used in bio-refineries. According to the ideal of cascading use of renewable material flows, they should also be examined for their suitability as high-quality building materials, as laminated timber or cross-laminated timber (CLT) made of poplar sticks. Such building materials could still be given to a bio-refinery (Fig. 4.13) after the end of the building's lifespan.

From the thin trunks of the approximately 4- to 8-year-old poplars rectangular rods can be cut, these then glued into boards and subsequently into laminated timber or cross-laminated timber. The process is similar to the production of CLT also as load-bearing components from soft and hardwood like the long-neglected beech, which—starting from European basic research—has been globally established in construction practice for a few years (Espinoza and Buehlmann 2018).

The poplar has special advantages for this project, which at the *Institute for Timber Construction* of the Biberach University of Applied Sciences in the years 2022 and 2023 was first examined in two final theses (Fig. 4.12): It is already approved as a raw material for glued products by European building material standards. The corresponding standard DIN EN 14,080 Sect. 6.2.2 allows gluing of "poplar (applicable clones: *populus x euramericana* cv "Robusta", "Dorskamp", "I214" and "I4551")"—and that as the only hardwood species. A building product made of poplar could therefore already be brought to market as standard-compliant today. Regarding the realisation of laminated timber and rod plywood from poplar sticks, however, some technical, economic and ecological questions arise (Schänzlin and Grossarth, 2022, see box).

Aspects that will decide the success of poplar CLT in construction.

(a) Manufacturing-relevant aspects
  Dimensions of the optimal cross-sections for manufacturing
  Optimized gluing process
  Formation of finger joints
  Achievable sorting class of the raw material
  Suitability of bio-based adhesives, such as those based on lignin
  Consistent consideration of the recyclability of poplar-CLT in terms of sustainability (further cascade use, energetic end use).

(b) Load-bearing relevant aspects
- Influence of the stick cuts on the strength
- How do branches affect the load-bearing capacity?
- Achievability of the strength of defect-free wood (according to DIN 68, 364).

(c) Economic aspects
- Optimal cross-sectional size of the logs considering the glue surface
- Consideration of the economic efficiency along the entire value chain from the producer (agriculture) through the processing (sawmill industry), building material production and trade to the end customer.
- Feedback to the production-relevant and load-bearing relevant aspects for optimising the cross-sections and the manufacturing processes.

(d) Sustainability aspects, such as agro-ecological. Under what conditions and in which areas can poplar plantations or poplar field edge structures contribute to the agricultural policy goals of increased biodiversity, increased soil fertility, and a decreasing need for plant protection products?

With regard to the resource issue, several cross-connections between agriculture and timber production are emerging that are being implemented in practice and accompanied by scientific research, such as agroforestry systems, in which trees provide shade or erosion protection, or the increased planting of field margins and tree strips, which are preferred for reasons of promoting biodiversity (breeding birds, beneficial insects) or also for wind and sun protection (water conservation in the soil by reducing evaporation). Poplar plantings are also suitable for this (Deutsche Bundesstiftung Umwelt, 2010). In any case, the proportion of materially used renewable raw materials from agricultural land can be greatly expanded (Fig. 4.13).

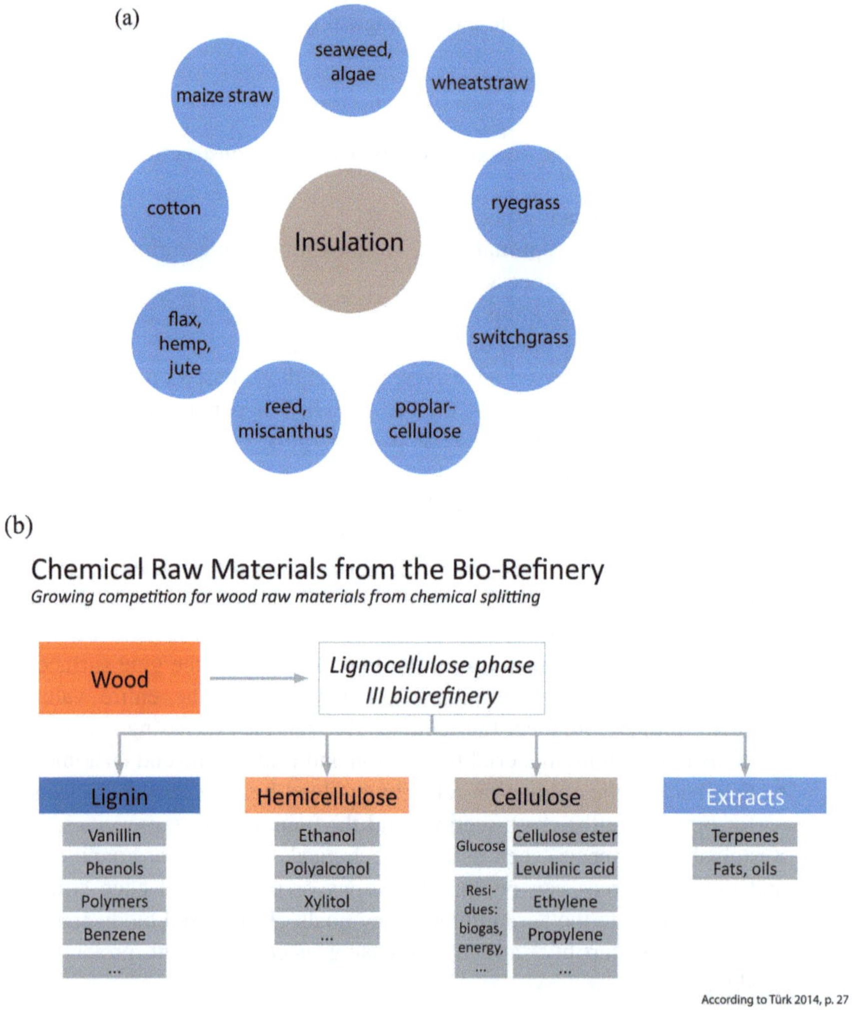

Fig. 4.13 Renewable raw materials and materials from a bio refinery. Sketch JG

Table 4.2 Research on CLT from hardwood since 2010. Selection JG (according to Espinoza and Buehlmann 2018, p. 4)

Country of study	*Examined wood type(s)*
Canada, Austria	Birch
China	Eucalyptus
Switzerland, Germany	Birch
USA	Hybrid poplar, American tulip tree
Germany	Birch and ash
Italy	Poplar and chestnut
Canada	Aspen, birch and maple

Hybrid-CLT Materials from Mixed Wood Species

CLT promises a higher-quality and also promising form of use for load-bearing components of the SRC woods. While so far mainly birch and beech have been the focus of corresponding research on CLT applications, also a few international research projects have examined the suitability of various poplar species for gluing CLT (Espinoza and Buehlmann 2018). Only a few research works examined the SRC-suitable poplar—especially the "Robusta" was researched here, sometimes combined with other woods such as walnut or chestnut (Table 4.2).

How Ecologically Sustainable Are SRC?

Short rotation coppice (SRC) are preferably established on low-yield sites or as part of landscape conservation or reclamation measures (Deutsche Bundesstiftung Umwelt, DBU, 2010). Unlike willows, additional fertilisation of poplars is generally not beneficial and therefore uncommon (ibid., p. 10). The cultivation of hybrid poplars is suitable for the renaturation of degraded riverbanks (Fortier et al., 2016). As perennial deep-rooters, they play a role in soil water management, for example by preventing flooding events after heavy rainfall.

Their leaf fall contributes to humus build-up in the order of 1 ton per hectare of SRC (DBU, 2010). The effects on biodiversity are positive when the SRC are limited to blocks of 1 ha each and green strips and paths are integrated (ibid., p. 27). Compared to arable use, poplar plantations often have ecological advantages (ibid., p. 54). In regional case studies, an economic advantage of agroforestry systems with poplar strips was demonstrated, such as for sorghum-wheat crop rotations in India (Chavan et al., 2022). While willows are more likely to be planted on arable land with better soil qualities and therefore compete more with food production, poplar cultivation in Europe often takes place on fallow or pasture land (cf. for Xu et al., 2021).

Over 100 years, the poplar (here: *populus spp. tacamahaca*) is estimated to be capable of storing less CO-equivalents in the wood through reforestation as part of the plans of the EU Commission and material use than willow and larch, but more than aspen and birch (Fig. 4.14). In this comparison the poplar reaches about 40% lower values than the willow. To arrive at these values, a very different planting density was calculated, which followed scientific recommendations on the optimal biomass production depending on the plant species. These were 1500 aspen and

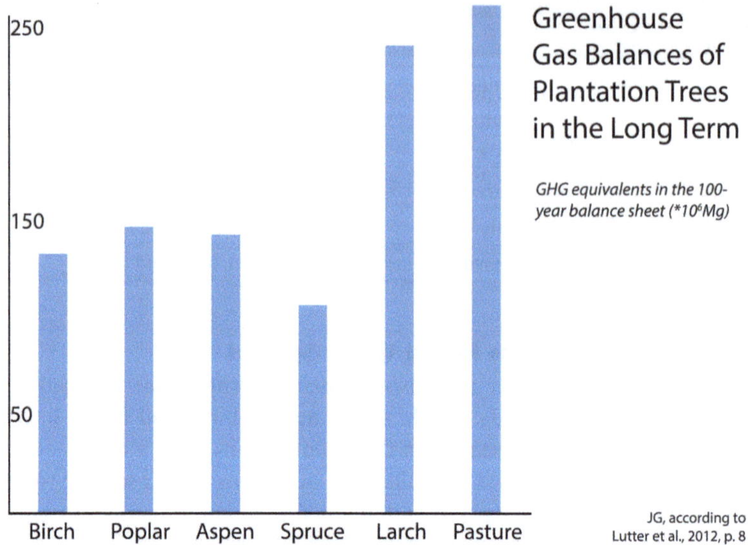

Fig. 4.14 Timber construction as a carbon store. Plant density or growth rate determine the long-term greenhouse gas balance of the cultivation of different tree species. In this comparison, willow would be preferable to other species. JG, according to Lutter et al. 2021, p. 8

poplar per hectare, 1600 birch and larch, 2000 spruce and 14,800 willow. For the interpretation of the numbers on CO_2 storage, it is but it is important to say that the poplar performed much worse than, for example, larches, because a larger proportion of energy use was assumed here (ibid., p. 11).

Aspects of an Economic Analysis of the Poplar

As far as the profitability of the poplar SRC is concerned, the harvest is the most significant cost factor in the life cycle of a poplar SRC. The same applies to the eco-balance with regard to the climate impact. This was the result of a life cycle cost analysis for Spain, which also makes the cultivation of poplars there for the production of wood chips unprofitable (Meyer et al., 2021; San Miguel et al., 2015). For Poland, field trials showed profitable cultivation (Stolarski et al., 2020). The focus in several current research projects is on increasing economic efficiency, where agricultural, plant breeding and environmental planning topics (*land use*) dominate (Landgraf et al., 2020). From a business perspective, according to a practitioner's report, the poplar even promises higher contribution margins than maize cultivation, higher harvest yields in terms of quantity than the willow, but its cultivation is considered more demanding.

Harvest quantities of 9–10 tons of agricultural dry matter, per year and hectare, were achieved (DBU, 2010, p. 50ff). The use of poplar as construction timber should increase the yield with the same effort. Newer hybrid poplar breeds are increasingly being used for reforestation—also in the form of forest SRC. They grow fast, quickly form shading crowns and get by with relatively little water

Fig. 4.15 The giant bamboo with its hollow interior is material-saving and tall-growing. It is not only interesting as a building material but also as an inspiration for lightweight construction. Photo Jan Grossarth

(Faust and Schirmer, 2022). The rotation time of the poplars is particularly low and is only about a third of the time an ash tree needs to grow. In forestry, conifers have so far been mainly planted for construction use, as the coniferous wood can be used more economically than hardwood due to its growth form, growth speed and properties. However, due to global warming, a conversion of the forest to hardwood is taking place. Forest trees grow until harvest for about 50–60 years (coniferous wood) or 80–100 years (hardwood). Due to the long time, there is a certain risk of a (total) loss e.g. due to storms or pest infestation. This is another advantage for the SRC trees, as the harvest sequence here is shorter.

4.7 Agricultural Renewable Resources

4.7.1 Grasses

Plant fibers from grasses can also be used for further processing in biorefineries. Grasses, like straw, have been and are being used in construction mainly as insulation materials. The biomass growth per hectare is high. Several harvests

Table 4.3 Heat storage capacity of selected marketable building materials: Specific heat capacity C (J/kgK) (Q. JG according to Holzmann et al. 2009, p. 16f)

Meadow grass	2200
Wood fiber	2000
Seagrass	2000
Grain granulates	1950
Hemp	1700
Flax	1640
Straw	1000
Concrete	880

per year are possible. Grasses can be used as cellulose "suppliers". Both ryegrass and miscanthus (Chinese reed) are currently mainly considered as "energy grasses". However, they can also be used primarily as blown-in or loose-fill insulation (Table 4.3). Miscanthus, for example, can be harvested with maize harvesting machines. It is undemanding in terms of soil quality and plant protection. A number of building materials such as insulation boards or infill bricks are already on the market, but the lack of building material standards also prevents widespread use here.

Ryegrass is often planted on animal pastures, but also in city parks or on sports fields. It is considered to be robust and fast-growing, reaching up to 90 cm. Ryegrass allows up to three harvests per summer season, so the fresh grass hectare yields can be up to 28 tonnes, which is more than three times that of wheat or corn. Dried, this results in about 8.5 tonnes, and from that about 4.25 tonnes of cellulose (Holzmann et al., 2009; p. 244f). The grass yields about 15% cellulose fibers after drying and processing. The remaining by-products can be used as protein feed, nitrogen fertiliser, or food additives.

The main products are usually still electricity and heat from grass biomass. Ryegrass pellets are also made into "bioplastic", which in turn can become a potential building material. Ryegrass is also considered the most suitable grass for the production of basic chemicals in so-called green biorefineries.

Compared to straw, the energy and climate balance is less positive. The reason is that the moistly mowed ryegrass must be dried before further processing.

For fire protection, the dried grass cellulose is treated with a borax solution (even though the European Commission classifies boric acid and borax as substances of concern under the REACH regulation as reprotoxic). Meadow grass insulation can be used as insulation between rafters, ceiling insulation, floor slab insulation, interior wall insulation behind cladding (DIN 4108), but also as biogenic reinforcement in cast concrete or in plasters. The heat storage capacity of the meadow grass is excellent, which qualifies it as insulation, but less so as support fibers (Table 4.3). The following sections provide, without claiming to be complete, an overview of the suitability of individual renewable raw materials from the field for construction purposes.

Ryegrass

Components: For construction purposes, the main component of interest is grass cellulose—the grass contains about 50 percent cellulose fibers. By-products are animal feed, amino acids.

Main use in construction: Interest in this mass-relevant natural fiber insulating material is currently increasing (Rüther, 2022). Dried for bulk or blown-in insulation; Building material class according to DIN 4102: B2. Also suitable for fleeces, sound insulation boards. Also Biopolymers based on grass fibers open up numerous application areas. For the production of the insulation boards, a binder is usually used. Cellulose reinforcement for plasters or concrete, or as injection molding polymer. "Due to their long, fiber-rich, firm leaves, which are traversed lengthwise by numerous parallel vascular bundles, the forage plant has been described as a fiber plant and raw material for paper production and for wickerwork already at the beginning of the twentieth century" (Holzmann et al. 2009, p. 246). Thermal conductivity λ: 0.040 W/(mK), bulk density ρ: 35–50 kg/m^3 (exposed), 40–65 kg (blown in), Water vapor diffusion resistance μ: 1–2 (Baunetzwissen 2023). Insulation in loose, blown-in or blown-on insulation. For example, as insulation between rafters, interior insulation, wall insulation in timber frame construction, ceiling or floor slab insulation. Meadow grass insulation made of 85% grass fiber pulp, 5% borax, 10% water. The borax improves fire safety. However, it is by the European Commission classified as of concern according to REACH regulation—namely: reproductive toxic—classified (boric acid and borax components).

Global cultivation area (ha): N/A.

Countries with the highest cultivation quantities: Ryegrass grows in large quantities in Germany itself.

Yield per hectare (t): 28 t per year with three harvests, resulting in 8.5 t of dry substrate and half as much grass cellulose.

Contributions to Ecologisation: Grass cover reduces compared with arable farming the soil erosion. Intensive cultivation with multiple annual nitrogen fertiliser applications, however, reduces the species richness of plants and insects. In arable farming, ryegrass is considered a "weed" and is chemically "fought" (Bayer, 2023).

Drought resistance: Increasing the resistance of the grass against drought stress is a goal of breeding research (LfL, 2022). It is to some extent salt tolerant (Marcar, 1987).

Recycling potential: If the insulation material is kept dry, it can be reused again and again. The flame-retardant treatment with borax reduces the organic compostability and reuse of the material as plant fertiliser. But there are already established procedures, that do without borax and work exclusively with fermentation processes, such as impregnations with whey and/or soda (FNR, 2009).

Overview

Life cycle assessment CO_2 equivalents: Preservation by silage (lactic acid fermentation) is not energy intensive. But washing and mechanical drying can worsen the energy balance and increase the Global Warming Potential (GWP).

Sources: Holzmann et al. (2009), p. 244ff; FNR 2009; Baunetzwissen 2023; LfL, 2023; UBA, 2022a, 2022b; THessen, 2023; Rüther (2022); LfL, 2022; Bayer, 2023; Marcar (1987).

Bamboo

Components: Bamboo (Fig. 4.15) contains about 68% cellulose and other carbohydrates, 25% lignin.

Main use in construction: Currently mostly in interior construction, for example for doors, floors, furniture, or as plastic composite materials for the outdoor area, for example terrace floors (Baustoffwissen, 2016). In Asia not only as scaffolding, but mainly for parquet production (from Moso Bamboo). Pressed as construction timber: planks, boards; insecticide resistance is traditionally also achieved by smoking or boiling. Strongly increasing importance as laminated timber or cross-laminated timber, for example in connection with the wood of the hemlock fir (Wie et al., 2019) or in other variants (Yan et al., 2022). The use of bamboo laminated timber (Glulam) in the structure, as beams, is likely to increase significantly. Here, there are challenges in wood processing technology (Kumar and Mandal, 2022). A current research concern is, for example, the question of how to improve the difficult adhesiveness of bamboo. Thus, the wettability and also the adhesive performance could be increased by bleaching or so-called plasma treatments (Nkeuwa et al. 2022). Bamboo is used in early stage experiments for bio-based 3D printing (Wong et al., 2024).

Global cultivation area (ha): In 2005, there were about 36 million hectares of bamboo forest worldwide. This was about 3.2% of the global forest area (FAO n.d., p. 12).

Countries with the highest cultivation quantities: India, Brazil, China, Nigeria, Indonesia.

Yield per hectare (t): Very different depending on the country of cultivation, type of bamboo, frequency of harvest. Between 4 and up to 26 tonnes of dry mass per year and hectare (NL Agency, 2013). Harvest for common varieties every 4 years (Moso Bamboo).

Contributions to Ecologisation: The harvest is still predominantly organized by small farmers. Bamboo is suitable for cultivating degraded land and improving soils (Solomon et al., 2020), which is being practiced as a model in many countries in Asia and Africa (Akoto et al., 2020; Dev et al., 2020).

Drought resistance: Bamboo usually requires moist growing conditions. Some species can better withstand longer dry phases according to dealer information

(Bambubatu, 2023). Genetic breeding methods are to be used to increase drought tolerance (Wu et al., 2020).

Life cycle assessment CO_2 equivalents: The rapid growth makes bamboo interesting as a construction material. Lower, equal or higher total carbon balances can be achieved per hectare of harvest area, depending on management, location and variety selection (van der Lugt et al., 2018). In particular, the possibility of using it for the reclamation of degraded land appears interesting.

Sources: Holzmann et al. (2009), pp.73–106; FAO (n.d.); van der Lugt et al. (2018); Bambubatu (2023); Wu et al. (2020); Baustoffwissen (2016): Dev et al. (2020); Akoto et al. (2020); Kumar and Mandal (2022); Nkeuwa et al. (2022); Yan et al. (2022); Wei et al. (2019); NL Agency (2013).

Reeds and Miscanthus

Components: The reed bed, which is also native to Germany, consists of 42–45% cellulose, 22–24% lignin. Miscanthus—not very widespread in cultivation yet, and if so, then for energy purposes, but also extensively researched as a building material—is suitable for use in the form of insulation boards, additives for lightweight plasters/concretes. The plants of the genus of perennial grasses, which typically grow between 1.5 and 3.5 m high, contain in the dry mass 35 to 50% cellulose, 10 to 25% lignin.

Main use in construction: Insulation boards are versatile in use. Thermal insulation, plaster-bearing interior finishing board, plaster carrier, thatched roofing; dowelled external thermal insulation composite systems (ETICS)—as in the case of all organic ETICS boards only above the splash water zone, usually three-layer with insulation boards, mineral base plaster, top plaster (Holzmann et al. 2009, p. 196). "In construction, reed has been used since the Early Stone Age for simple tent-like dwellings, later roof coverings, and interwoven with hazel or willow rods as a stabilizing supporting structure (plaster carrier) for clay or lime plasters." (Holzmann et al. 2009, p. 185) Miscanthus analogue, but also as a concrete additive, as a basis for paper or biopolymer production.

Global cultivation area (ha): N/A.

Countries with the highest cultivation quantities: Germany, Great Britain, China, India.

Yield per hectare (t): In Germany an average of 12–20 tonnes per year; peak values in Europe of up to 30 tonnes. Reed grass: 8 to 15 tonnes of dry biomass (Herrmann et al. 2016; Lewandowski et al. 1999). 8–11 t (sandy soils) up to 25 t.

Contributions to Ecologisation: Miscanthus can also be grown in agroforestry and agroecological cultivation systems. Due to its deep roots and its ability to store carbon and nutrients in the soil, it has potential to improve soil fertility and to reduce erosion. Reed grasses are supposed to be grown and decomposed as a result of peatland rewetting and economically valorized. Miscanthus is drought-resistant and can be grown in regions affected by droughts or dry periods. Miscanthus has a deep root system that reaches up to three metres deep into the soil and allows it to use water from deeper soil layers.

Sources: Holzmann et al. (2009), pp. 182–214; p. 260ff.; Herrmann, C., Idler, C., & Heiermann, M. Biomass from perennial crops—a sustainable source of renewable energy. Journal of Cleaner Production, 127, pp. 332–341.: Lewandowski I, Kicherer A, von Braun J (1999). A comparison of the energy yield of reed canary grass, wheat and winter barley grown on the same area and under different conditions in terms of life cycle analysis. European Journal of Agronomy, 10(2), pp. 95–105.

4.7.2 Grains

Grain is used as food and feed. But its grist was and is also used as insulation: As grain insulation filling in the form of pour, single or blow-in insulation, between rafter insulation, interior insulation, wall insulation in timber frame construction, ceiling or floor panel insulation; cavity insulation of bricks; Grain insulation (approximately 35% rye bran, 35% rye husk, 25% hydrated lime, 4% potassium water glass, 1% whey).

Grain bran only burns when directly ignited (Holzmann et al., 2009, p. 121ff.) Grain straw, like hemp, is a versatile building material. In addition to its use as insulation, this agricultural raw material has also been tested as a load-bearing wall material—plastered with clay, straw bales as impact sound insulation panels. Straw bale houses were first built in the United States in the late nineteenth century due to mechanical pressing techniques, the first in Europe has been in France for over 100 years. As well as rye or wheat straw, hemp straw is also suitable.

Grain straw is of particular importance for construction. Wheat, spelt, rye straw, but also straw from the rice plant or from triticale are mainly used in straw house construction (BauStroh GmbH). In addition to its excellent greenhouse gas balances, the main advantage of construction straw is its excellent heat storage capacity, as well as high sound absorption values (Ehling, 2020). Solidly built straw bale houses and also straw-timber frame houses achieve U-values of passive houses (Bernard et al., 2014). This innovative application field is economically favorable: timber frame straw modules can be partially standardized in advance ("Mod-Cell panels"; Bocco, 2014).

In addition to the wall construction options, straw bales can also be used for the construction of roof structures. They should be used in a thickness of only 36 cm. They are also used as ground foundation insulation, for moisture protection on elevated screw foundations.

Table 4.4 Grain and straw yield of selected cultivated species (Holzmann, 2006, p. 160)

Winter wheat grain	30-55dt/ha	straw 45-75dt/ha
Winter rye grain	25–50 dt/ha	straw 50–80 dt/ha
Oats grain	30–50 dt/ha	straw 40–55 dt/ha

The energy balance of straw insulation is far superior to that of petroleum-based insulation. For the production of a straw wall insulation with the U-value 0.10 W/ $(m^2 \cdot K)$, only about 10 kWh/m^2 are stated. A common XPS insulation, on the other hand, comes in at more than thirty times the value, 345 kWh/m^2. A wood fiber insulation board with a similarly high value of 329 kWh/m^2 (Bernard et al., 2014).

Straw is also used in clay bricks. It can contain proportions of about 20–30%. However, sawdust additives are also common as "renewable alternatives" (Illampas et al., 2014).

For clay plasters in the style of traditional half-timbered houses, straw is known to be a bonding base, with thin, long straw being preferred—i.e., just the kind that is no longer "available" in conventional agricultural cultivation, which uses hybrid grains with very short stalks.

A mixture of straw and clay, known as "well clay", is on the market for "alternative building materials". Its production—interesting from the point of view of raw material availability—requires thin straw stalks of 30–40 cm in length (Schäfer, 2021). For the economic consideration, the agricultural yields of various grains are interesting information. In terms of stalk yields, rye is generally preferable to wheat or oats, although it depends on the specific location (Table 4.4).

4.7.3 Additional and Marine Sources

Numerous other renewable raw materials can at least be used as insulation material. These include algae (micro- and macro-algae), seagrass (as insulation in loose fill insulation or blown-in insulation). Here additives of (toxic) borax for fire protection is not necessary due to the naturally high salt content (Holzmann et al., 2009, pp. 215–226).

According to general information, between 600 and 1200 m^3 of seagrass are washed up on the German coasts, depending on the region. Since washed-up seagrass may only be disposed of with appropriate pretreatment and the spreading of the washed-up, untreated plant parts on the fields is prohibited in parts of Germany, this beach litter often becomes a regional waste disposal problem. However, one can expect that this will change with the rediscovery of the plant raw material seagrass and the known general environmental situation (ibid., p. 218).

Also suitable are: flax, cork (e.g. as insulation boards for ETICS), cotton—also recycled from used clothes like jeans, from which insulation materials are made—, coconut, pineapple fiber (cultivation as fiber plant without fruit sand for example in Taiwan, Philippines—great robustness suggests suitability as insulation fiber

Fig. 4.16 A fat from insect biomass in a laboratory of ETH Zurich. Photo Jan Grossarth

raw material (Holzmann et al., 2009, p. 255). Jute (in Germany 1900 still with a cultivation area of around 1 million hectares (Holzmann, p. 258), hops, which historically was also used as a fiber plant, nettle, which was also traditionally used as a fiber plant for textile production. The intermediate stage of the insect, which feeds on agricultural residues, can produce not only edible fats but also lipids, which can be used as lubricants or for wood processing (Fig. 4.16).

4.8 Agroecology as a Policy Concept

In view of the ecological damage caused by large-scale monocultures in "cleared" agricultural landscapes, renewable raw materials should be provided against the background of agroecological farming systems. But what is agroecology? It cannot be defined in one sentence. Perhaps because of its complexity it is suitable as a political guiding perspective, so to speak "to please everyone". The European Commission (2019), for example, relies on the concept in the context of its "transformation" ten-year plan "Green Deal", changed into the "Clean Industrial Deal" in 2025. Also, the food policy strategy of the EU Commission with the name "Farm to Fork" — canceled in 2025 — refers to it. The World Food Organisation FAO has its highest experts from the Advisory Committee on World Food Security (CFS) illuminate the concept on 163 pages (*HLPE*-Report, 2019). The summary alone requires eleven bullet points for a definition. Agroecology means therein a preference for.

> natural processes, it limits the use of external inputs, supports closed cycles with minimal negative external effects, emphasizes the importance of local knowledge and participatory processes

and should also reduce social inequality and help the sciences to gain recognition.

This complexity, however, also carries the risk of vagueness. So this, possibly: a container word? Agroecology pleases, colloquially speaking, perhaps just because it is less rigidly defined than the many "organic" standards. And because, due to its approach of adapting agricultural cultivation to local conditions, it appears more practicable than the path of increasing yields through agro-industrial intensification modelled on the 'green' revolution of the North, which has been recommended for Africa by international organisations for decades. This meant, from the 1950s onwards, an agrochemical path. This not only had health and ecological side effects, but often failed due to lack. For how can intensification succeed if farmers lack money, fertilisers, and pesticides?

Agroecology aims to intensify agriculture through a so-called enhancement of ecosystem services, but also "conventional" measures such as demand-oriented fertilisation are fundamentally allowed. The integration of local farmers' experiences and science-based recommendations that concern the local situation are emphasized. According to the *HLPE*—Report, this is a "dynamic concept", a "transdisciplinary science", promoting "biological-ecological processes" in food systems; agroecology is: inclusion of local experiential knowledge, a science, a social movement, an "innovative approach to "food systems" (HLPE). The concept is clearly distinguished from "sustainable intensification" (HLPE, p. 63). Agroecology is therefore labor-intensive rather than capital-intensive, is comprehensive with political-social aspects such as democratisation, "justice", human rights. On the agronomic level, it is thus nutrient-efficient and biodiversity-friendly. According to the HLPE-Report, it includes among other things: "nutrient cycles, biological nitrogen fixation, improvement of soil health, water protection, biodiversity protection, habitat management, carbon sequestration, biological plant protection, diversification in arable farming [...] compost use" (ibid., p. 36).

The 13 principles of agroecology defined by the HLPE are:

1. Recycling, predominantly use of local nutrients and building materials
2. Reduction of off-farm inputs (fertilisers, chemicals)
3. Improvement of soil health
4. Improvement of animal health and welfare
5. Increase in biodiversity
6. Utilisation of ecosystem synergy effects
7. Economic diversification (broadening the income base of small farmers)

8. Co-creation of knowledge: Improvement of the co-creation and horizontal dissemination of knowledge, taking into account local and scientific innovations
9. Social values and diets: Building food systems based on culture, identity, tradition, social and gender justice
10. Fair prices for farmers, preservation of property rights
11. Connectivity. As direct contact and short distances from producer to consumer as possible
12. Governance of land and natural resources
13. Participation: Social organisations and decentralised government agencies participate in the value chain.

This list also makes it clear why, on the one hand, almost every interest group finds a point somewhere that suits their interests. Some points prove to be empty formulas upon closer inspection. For what exactly should be understood by "fair prices" be? And on the other hand, it becomes clear what the "spirit" of agroecology is: subsidiarity; consideration of local cultural interests; respect for experiential knowledge. This approach is also popular in transdisciplinary research, which includes local agricultural practice.

An example of an approach accompanied by agricultural science is an intensified "rediscovery" of traditional cultivation methods of cereals in crop rotation and in proximity to other field crops (multicropping) and their improvement. The so-called multicropping of peanuts and millet, for example, is accompanied by scientists in terms of many factors such as productivity and impacts on the local soil ecosystem.

Agroecological fish farming: The "Fish CRP" example.
A concrete example is "Fish CRP", an initiative of the CGIAR's fish research program in many African and Asian states, behind which are the international research centre World Fish, the Dutch University of Waageningen, the Australian James Cook University and the University of Greenwich (Li et al., 2022). In villages, small fish farms are created where farmers can "refine" grain into fish. This does not consume drinking water, but is based on groundwater or takes place in rivers, lakes and ponds. Plant waste becomes fish food. The dirty water serves as fertiliser for the fields. There, in the next step, ideally more than just corn grows, but a variety of fruits and in between exist push and pull cultures that attract beneficial insects or distract pests, thus enabling pesticide-free agriculture.

Agroecology is therefore a vague concept, or more benevolently said: a versatile one. One danger is on the one hand that the richness of facets invites cherry picking. On the other hand, it connects—similar to the bioeconomy—many different actor groups. In principle, it also leaves the "door open" for the use of pesticides and nitrogen fertilisers, if they are "selective" (HLPE-Report, among others p. 37, p. 46, p. 81). Organic farming as a defined standard has large overlaps with agroecology. But this can be understood not only as a way to it, but also "beyond" it. Because it is interested in *socio-cultural* aspects of land use ("connectivity", "fairness").

For agroecology, there are already support programs for cultivation in individual states, such as Brazil, or even state food purchase programs in a socio-political context. However, there are (still) no broadly developed markets that reward the special approach, as in organic farming. In this sense, agroecology still occupies ministries and even more scientists who want to look beyond the horizon of their departments and disciplines. It also has this in common with the bioeconomy and the circular economy.

4.9 Land Use Competition and Solutions

The bioeconomic path of resource replacement intensifies the desire for land. If the climate policy plans and the expected, necessary expansion of bioeconomic resource use by the federal and state governments were realised, the demand for biomass would increase sharply. The demand for woody raw materials for material use would already increase by 0.5% annually in a forecast scenario that assumes moderate increases in demand in Germany. The demand for biomass by the chemical industry would even increase by 5.1% annually (Poganietz et al., 2020, p. 194).

The main ethical dilemma of using renewable resources for construction is the competition for land with food. On the one hand, "Food First" is considered the ethical consensus of the bioeconomy. On the other hand, in international development policy, more and more an approach that focuses on a socio-ecological "optimization" of the entire so-called *Food Systems* (von Braun et al., 2023). Here, the efficiency of agriculture is as much a concern as the health of people and the climatic consequences of farming. This new focus essentially promises to no longer think of the "opposition" of land use as rigidly as before. Two examples: The material use of wood permanently removes CO_2 from the earth's atmosphere, which can have a positive effect on climate stability. And a reduction in the area-consuming animal husbandry would be able to defuse the competition for land even with a growing world population. The issue of human malnutrition is closely linked to inefficient land use. For decades, the world population and the proportion of overweight people have been increasing equally, and the latter also in developing and emerging countries (Grossarth, 2019).

Systemic networking is required. In particular, there is scientific evidence for greater area productivity in biomass production through agroforestry systems. Nevertheless, higher management costs are mentioned as an obstacle to implementation (Lehmann et al., 2020). Agroforestry can be particularly interesting for timber construction, as numerous woody plants can be integrated, such as hazelnuts, bird cherries, walnut, service tree or various fruit trees. In field trials in arid areas with degraded soils, significant improvements in soil quality, such as water storage capacity, and an increase in yields were demonstrated (Ikromov and Kadamov, 2021).

An agricultural approach to reducing the ecological conflicts of land use could also be the increased use of perennial crops, because on average, the raw material efficiency in relation to ecological and intensive cultivation is relatively more favorable in the former—that is, they also come closer to the yields of intensive agriculture with less energy-intensive fertilisation and pesticide use than annual plants. Cassava in Africa is an example. Miscanthus, which provides building materials and energy, is an example from Central Europe (Fig. 5.3). So many aspects need to be considered when the greatly increased use of renewable raw materials in construction is demanded, but so great is also the breadth of creative designs of landscape planning and use.

Important levers are also (once the question of how democratically legitimized it would be to "turn" them is excluded):

Meat consumption: It has already been described that large parts of the agricultural areas are used for the production of animal proteins (Fig. 4.17). Wouldn't it be necessary to ask here, in view of the climate political realities, whether timber construction (i.e., climate protection from the area) should be preferred to intensive feed production? It would probably be the more obvious question than to discuss "houses or plates?" here.

- *Biomass for power generation:* In addition to high meat consumption, the high land requirement of wood chip power plants or biomass plants should also be considered. Solar energy and wind power plants generate fifty to one hundred times the electricity yield in kilowatt hours per hectare of land. No other form of energy generation exacerbates the competition for land as much as biomass (Dumke, 2020, Tab. 41f.). Biomass plants that are "fed" with straw generate an energy yield of less than with a quarter of the same cultivation area, a much higher area requirement for power generation than the fast-growing plants Miscanthus or corn (ibid., Tab. 28). This is a strong argument for building with straw and not fermenting it, and also not burning it.
- *A circular economy of plant nutrients:* Already today, agricultural yields depend to a large extent on the supply of artificially produced fertilisers such as nitrogen, phosphorus, and potassium. The yield advantages of European or North American agriculture compared to African agriculture can largely be explained by the different fertiliser expenditures (Grossarth, 2019). After all, the agricultural productivity increases since the beginning of industrialisation were predominantly subsidized by
- Mineral fertiliser (organic like guano, later chemically synthesised nitrogen) from the 1840s
- Oil, later natural gas to N-fertiliser (Haber–Bosch process, from 1910s)
- Mechanisation (replacement of the horse, the ox by tractor), widespread in Western Europe from the 1950s
- Agrochemistry (plant protection), from the 1960s.

Fig. 4.17 Biggest competitor for the area where agricultural building materials could also grow: the gluttonous livestock. Photo Jan Grossarth

Land use conflicts can also be mitigated by a multiple use or multiplication of the area, for example by so-called *multi-cropping*, agroforestry, vertical farms in high-rise buildings or by agricultural PV production systems on fields. In the latter case, the partially transparent PV roofs still allow a—usually slightly reduced—growth of grain, apples or grapes above the ground or at shrub height. In particular, the methods and approaches of agroecology and agroforestry are trusted to mitigate the land use conflicts of the bioeconomy (Table 4.5). These are soil improvement, which is based on a humus build-up *carbon farming*) and agroforestry. Possible "*win–win*-effects" of the so-called *carbon farming* or agroforestry are outlined in the following table.

Finally, the above-outlined "biotechnological revolution" of this decade is also associated with hopes that the land competition of the bioeconomy can be reduced. The "disruptive" developments also include precision fermentation, the production of "cheap proteins" in the bioreactor from grass or CO_2 (Vanhercke and Colgrave, 2022). The drastically falling costs of key technologies such as genome sequencing or the computing capacity of computers are considered drivers.

Table 4.5 Evaluation matrix, JG, IPCC, Lampking 2015, Pearce 2015

Agricultural approach	Productivity	Use of non-renewable energies, greenhouse gas emissions	Biodiversity and associated ecosystem services	Soil quality and water resources	Profitability
Carbon increase in the soil	+ (better than in conventional agriculture)	+	++ (much better than in conventional farming)	+	0 (similar to conventional farming)
Reduced use of synthetic pesticides and fertilisers	– (significantly lower than in conventional farming)	+	++	+ +	–
Agroforestry	+	++	++	++	± (depending on location, yield for timber)

Case Study: Agrivoltaics and Farmland

In 1981, a contribution by the German scientists Adolf Götzberger and Armin Zastrow appeared in the magazine "Solar Energy". It was titled "Potatoes under the collector" (Götzberger and Zastrow 1981; Fig. 4.18). The authors suggested that fields should be used in such a way that two things would be harvested there: potatoes and solar power. Because the potato prefers the shade of the photovoltaic modules (PV) to the zenith sunshine. The idea of "Agri-PV" was born. It then took more than forty years for the idea to also excite the major politics. On July 7, 2022, the Bundestag passed an amendment to the EEG law. It states that Agri-PV is now also allowed on grassland. And the Federal Ministry of Agriculture announced that the EU area premiums, which are vital for farmers, would continue to be paid for the combination of collector, potato or grassland. This means: Agri-PV is also legally considered agriculture, not power plant construction.

Neuer Vorschlag der Fraunhofer-Gesellschaft

Kartoffeln unter dem Kollektor

Solarfarmen und Landwirtschaft müssen nicht unbedingt in einem Gegensatz zueinander stehen. Beide Bereiche lassen sich vielmehr vortrefflich miteinander verbinden. Daß das funktioniert, zeigen beispielsweise entsprechende Versuche in Saudi-Arabien.

Neu ist jetzt ein Vorschlag für eine besonders günstige Anordnung für Solarenergieanlagen in Verbindung mit der landwirtschaftlichen Nutzung. Dieser Vorschlag wurde von Prof. Adolf Goetzberger und Armin Zastrow vom Institut für Solare Energiesysteme in Freiburg erarbeitet. Der Trick dabei ist, daß die Kollektoren nicht direkt über dem Boden, sondern in etwa zwei Meter Höhe angebracht werden. Wenn dann der Abstand zwischen den Kollektorreihen etwa das Dreifache der Kollektorhöhe beträgt, erhält man eine fast gleichmäßige Bestrahlung des Bodens. Der Pflanzenbau wird dadurch nicht beeinträchtigt, denn die Photosynthese läuft bei diffusem Licht besser ab als bei direktem Sonnenlicht.

Beide Wissenschaftler berechneten den direkten und den diffusen Anteil der solaren Strahlung und zeigen die Vorteile einer solchen neuen Anordnung auf. Sie wollen gleichzeitig zeigen, daß die Vorurteile über die Landschaftszerstörung und die angeblich lebensfeindlichen Auswirkungen größerer Solaranlagen einer ernsthaften Nachprüfung nicht standhalten.

- Um optimale Energiekonversion zu erreichen, muß jeder Flachkollektor nach Süden (auf der Nordhalbkugel) ausgerichtet werden und um einen Winkel, der gleich der geographischen Breite oder etwas größer ist, gegen die Horizontale geneigt werden (4).
- Um zu starke gegenseitige Abschattung zu vermeiden, sollen die Kollektorreihen einen gewissen Abstand haben, der üblicherweise gleich der dreifachen Kollektorhöhe gewählt wird.
- Aufgrund dieser Anordnung erreicht ein großer Bruchteil der solaren Strahlung den Boden zwischen den Kollektorplatten. Dieser Betrag wird insbesondere in den Sommermonaten während der Wachstumsperiode groß sein.
- Wenn die Kollektoren direkt auf den Boden gestellt werden, wird die Strahlungsdichte zwischen den Kollektoren sehr inhomogen sein, wobei direkt unter den Kollektoren fast nur Schatten zu finden ist.
- Durch Anhebung der Kollektoren auf ein geeignetes Gerüst kann die Strahlungsverteilung ausgeglichen werden.
- Solarkollektoren werden auf gute Lei-

Grundlagen der Rechnung

Die mathematischen Beziehungen, die für diese Berechnungen hergeleitet wurden, basieren auf folgenden Annahmen:

1. Das periodische Kollektorfeld ist nach Süden ausgerichtet und zweidimensional unendlich ausgedehnt.
2. Die Stützen, auf denen die Kollektoren montiert sind, tragen nicht zur Abschattung bei.
3. Die Kollektoren sind einschließlich ihrer Rückseiten ideale Absorber. (Das ist nicht notwendigerweise so, siehe unten.)
4. Die diffuse Strahlung ist isotrop bezüg-

Das Fraunhofer-Institut für Solare Energiesysteme in Freiburg

Dieses neue Institut der Fraunhofer-Gesellschaft begann am 1. Juli 1981 mit seiner Arbeit. Es ging aus einer Arbeitsgruppe des Fraunhofer-Institutes für angewandte Festkörperphysik hervor,

Fig. 4.18 Pioneering? The idea of Agri-PV was born in 1981 with this article, but only considered politically important from 2022 onwards. Picture quote Solar Energy

Two important levers have been adjusted. But what do the farmers actually say? They have reason to think about Agri-PV: The drought is pressing, shade donors are vital. Even in the Bavarian Rhön, the soils are dusting again. Here too, the question hovers above everything: How will we continue to feed ourselves from such desert land in the future? Together with his son, farmer Mathias Klöffel cultivates 180 ha of land in Großbardorf in the Rhön. Wheat, corn and other crops ripen here. Klöffel has already done a lot to adapt agriculture to the new climate conditions. Moving away from corn, towards crop rotations with triticale, deep-rooted alfalfa, legumes, sage. Much of this goes into the biogas plant. He doesn't seriously consider Agri-PV.

Klöffel would be an ideal candidate for Agri-PV. Wouldn't it be nice if panels protected his fields from drying out quickly? Klöffel, also the managing director of the energy cooperative Agrokraft, smiles. He mainly thinks about the extra effort. He calculates using his business as an example: a hectare of wheat land brings 300 euros into his operating cash register, and that's not exactly a lot. The work has to be uncomplicated. "If there are five

PV systems in between," he says, "the working time for mowing becomes much too long."

His personal path into solar energy was less complicated: He had his barn roofs covered with PV modules 20 years ago. And the cooperative operates a classic solar system that takes up space. But concerns, like the Klöffels', could one day be dispelled. Agri-PV technology is evolving. Large tractors and combine harvesters could drive through the rows - the height of the PV roofs is 6 m, the distance between the panel rows is 18 m. In Geisenheim in the Hessian Rheingau, the harvest should be under Agri-PV roofs in 2023. Not grain, but grapes. Three metres high, solar roofs cover the young vines here. The area is small: half a hectare. (HS Geisenheim, 2020).

But beyond such research projects, in practical agriculture, Agri-PV is still largely sought in vain in Germany. Markus Haastert wants to change that. The entrepreneur in Berlin, who has also practically accompanied some research projects, is betting on this field with his *Agro Solar Europe GmbH*. He advises farmers and reports of greatly increased interest after the parliamentary decision in July. "The topic is gaining momentum," he says. His company has provided consultations for 400 ha of land in the past few weeks alone; this is still a relatively tiny area. In Haastert's consulting business, it's about questions like the optimal shading of fruit in the Altes Land, berries in Brandenburg, potatoes and beets in the Rhineland. But it's also about more natural facilities, says Haastert. "We don't want to pour concrete on the field, but facilities on filigree steel anchors that are made like tree roots."

But the ball is still in the field of science. Many research questions are open: How can panels be set up to provide hail protection in spring, but also improve the microclimate in summer? How can irrigation systems be integrated? What are the effects on biodiversity? What are the environmental impacts on breeding birds—for example, due to glare effects? What are the specific figures on the water balance of Agri-PV? In agricultural science, it is often, and also in this case, that meaningful answers can only be given in very specific application contexts. Each crop rotation, each soil, each climate region has its own needs. Accordingly, the list of ongoing Fraunhofer research is colorful: It's about the optimal shading of grain in Chile, the use for local supply in Mali and Gambia, solar roofs for shrimp farms in Vietnam. But Science already knows a lot today about which application fields are agriculturally sensible. Permanent grassland or cereals can tolerate shading in the morning and evening. This can be achieved through ground-level, vertically positioned modules. The southern German energy supplier *Lechwerke* has been experimenting with this for about a year in the Allgäu. These panels allow for grazing by sheep, goats, or chickens.

Fruit crops such as pears and apples or potatoes, which need sun protection at noon, can rather be combined with roof-shaped PV modules on stilts. So far, there are only a few experimental facilities in Germany—one at Lake

Constance, one in Rhineland-Palatinate, where Demeter apple orchards are successfully protected from sunburn.

At the Freiburg ISH Institute, Max Trommsdorff heads the research on Agri-PV. In his eyes, their time has now come, and "faster than thought". He sees, as he says, in Agri-PV "a technology that has the potential to reduce water consumption and make land use much more efficient". A lot is also happening in material research: increased interest in so-called organic PV films, or thin-film PV, which precisely let through the light spectral colors required by the respective cultivated plants. "There is also a trend towards so-called bifacial PV systems, the back of which is also electrically active," says Trommsdorff. This results in higher electricity yields per square metre of module area (Trommsdorff et al., 2021). Trommsdorff has not only traveled the world as a guitarist and written a book about it. He is also an economist, and he knows that the implementation in practice is decided not only by good scientific arguments but primarily by business economics. Now the farmers have clarity: the EU agricultural subsidy continues if there is a "main agricultural use", i.e., no more than 15% of the usual yield is lost due to the solar systems. Clear criteria for animal husbandry are still lacking, says Trommsdorff. But how much more expensive is a high-standing PV system, like the one in Heggelbach? The competence centre for renewable raw materials Straubing of the Bavarian state government has documented data on this. 1234 euros per kilowatt-peak of electricity yield are incurred for high-standing systems, instead of only 572 euros for common open-space systems. Because cultivation is possible under the more expensive solar roof, the income from the sale of grain or fruit is added, as well as the approximately 300 euros EU area premium. These figures show that Agri-PV will initially be profitable on very productive soils, where it is lucrative to receive harvest revenues. In addition, the fact that solar systems are becoming increasingly productive. ISH researcher Max Trommsdorf recalls the decline in feed-in tariffs due to the EEG levies, he says: From more than 50 cents (2000) to around 5 cents per kilowatt hour currently, this has fallen. It is hardly a factor for investors anymore. The consultant Markus Haastert from Agro Solar Europe says: "Most farmers today decide to bypass the EEG and sell the electricity directly to customers from the to market the regional industry."

Agri-PV is just one of several possibilities of so-called integrated photovoltaics. These are all the fields where solar panels are built into existing infrastructures: from house facades to rail networks, over car roofs to gravel pits. The expansion potential of Agri-PV is particularly large here, ISE researcher Trommsdorff has calculated with other Fraunhofer researchers. For this, he compared the potentials of Agri-PV with other fields: They came up with 1700 gigawatts-peak, which is significantly more than for the solar expansion of buildings, roads and rail routes, water surfaces, noise barriers or a corresponding use of car surfaces. And it would be many times more than would be necessary for climate neutrality in the energy sector according to calculations by the ISE. But the expansion does not go with the sledgehammer. There are conflicts with the landscape aesthetics. For this Fraunhofer ISE also conducts its own acceptance research. First results: Agri-PV is perceived by the population as less disturbing than pure PV areas or especially wind parks. Also for one of the most common field crops, the technology is hardly suitable. The corn, which as a so-called C-4 plant particularly effectively converts light into biomass, is considered unsuitable for PV combinations. For corn it is ideal to have as much sunlight as possible. It can't get enough of photosynthesis.

4.10 Summary and Outlook

The use of renewable raw materials in construction is likely to increase with historically unprecedented intensity against the backdrop of global climate protection policy. The traditional consideration of "solid wood" as the main building material requires a strong expansion—especially towards agricultural raw materials and their processing using innovative biotechnologies.

Many biogenic raw materials are or will become relevant to the construction industry as bulk building materials: straw and grasses, such as ryegrass—and bamboo in Asian and African regions—, flax or cotton, coconut, pineapple, reed, cattail, hops, jute, nettle, paulownia trees, sugarcane.

The use of renewable building materials is in the context of an already large and increasing land competition, with the political approach of agroecology, with the priority of food production and in relation to the chance of an increasing high-quality recycling of organic waste (colloquially: waste streams). Every innovation towards a circular economy reduces the "pressure on the area", i.e. the need for renewable resources. Therefore, a bioeconomy is necessarily embedded in the framework of a circular economy.

Bioeconomic construction in this sense combines traditional with innovative raw material (usage types). Since the biogenic construction represents a cultural break with a construction practice that has been focused on demolition and resource-intensive new construction for at least seventy years, not only material science innovations and efficiency advances in the agricultural-forestry raw material production are necessary, but also *Value-Chain* innovations, as they were addressed in the introductory chapter. The beauty and the pleasant indoor climate of, for example, clay-plastered straw bale houses can also invite people to perceive even smaller living spaces as worth living. Such cultural-lifeworld references easily get lost from the technical perspective on "Renewable raw materials and biotechnologies". Since the "transformation" would also mean cultural breaks in this field, resistances and reactions from actors such as livestock farmers are to be expected, which set limits to the environmental planning perspective. Their interests must be seriously considered.

References

NL Agency. (2013). Bamboo. Analyzing the potential of bamboo feedstock for the biobased economy., n.p.

Aitken, S. N. et al. (2008). Adaptation, migration or extirpation: Climate change outcomes for tree populations. *Evolutionary applications, 1*(1), 95–111.

Akoto, D., et al. (2020). Towards bamboo agroforestry development in Ghana: Evaluation of crop performance, soil properties and economic benefit. *Agroforestry Systems, 94*, 1759–1780.

Americanhardwood. (2023). *Maggies Oldham.* Accessed on 10.10.2022. https://www.americanhardwood.org/en/examples/our-projects/maggies-oldham/outcome

Amziane, S., & Sonebi, M. (2016). Overview on biobased building material made with plant aggregate. *RILEM Technical Letters, 1*, 31–38.

Bambubatu. (2023). *Drought tolerant bamboo.* Accessed on 7.9.2022. https://bambubatu.com/drought-tolerant-bamboo-for-dry-climates/

Bardi, U. (2009). Peak oil: The four stages of a new idea. *Energy, 34*(3), 323–326.

Bartsch, N. & Röhrig, E. (2015). *Waldökologie: Einführung für Mitteleuropa.* Springer.

Bastin, J.-F., et al. (2019). The global tree restoration potential. *Science, 365*(6448), 76–79.

Baunetzwissen. (2023). *Wiesengras.* Accessed on 12.8.2022. https://www.baunetzwissen.de/daemmstoffe/fachwissen/daemmstoffe/wiesengras-746408

Baustoffwissen. (2016). *Bambus.* Accessed on 4.9.2022. https://www.baustoffwissen.de/baustoffe/baustoffknowhow/energetisches-bauen/bambus-baustoffe-parkett-terrasse-fassadendiele-furniere-innentueren/

Bayer. (2023). *Weidelgras.* Accessed on 5.11.2022. https://agrar.bayer.de/Agrar%20Magazin/Weidelgras_Verbreitung_in_Deutschland

Bernard, T., et al. (2014). Bautechnische, ökologische und ökonomische Grundlagen der Planung von Strohballen-Gebäuden—Literaturzusammenstellung und Analyse des thermischen Leitwertes der Gebäudehülle. *Bauphysik, 36*(3), 134–143.

Bewg, W. P., et al. (2018). Genome editing in trees: From multiple repair pathways to long-term stability. *Frontiers in Plant Science, 9*, 1732.

Bocco, G. A. (2014). Architect werner schmidt's straw-bale construction, non-conventional materials and technologies for sustainable. *Engineering, 600*, 727–738.

von Braun, J. et al. (2023). Science for transformation of food systems: opportunities for the UN Food systems summit. *Science and Innovations for Food Systems Transformation*, 921.

Bringezu, S. et al. (2020). *Pilotbericht zum Monitoring der deutschen Bioökonomie.* CESR, Kassel.

Bundesministerium für Ernährung und Landwirtschaft BMEL. (2019). Ackerbaustrategie 2035. Accessed on 18.8.2022. https://www.bmel.de/SharedDocs/Downloads/DE/Broschueren/ackerbaustrategie2035.pdf?__blob=publicationFile&v=8. Last checked on 18.08.2022.

Bundesumweltministerium BMU. (2012). *Langfristszenarien und Strategien für den Ausbau der erneuerbaren Energien in Deutschland bei Berücksichtigung der Entwicklung in Europa und global, Berlin.* https://www.erneuerbare-energien.de/EE/Redaktion/DE/Downloads/Studien/studie-langfristszenarien.pdf;jsessionid=63149C0BDA12F980714FF00F860EB7A0?__blob=publicationFile&v=4. Last checked on 18.08.2022.

Calder, B. (2021). *Architecture: From prehistory to climate emergency.*

Cao, S., Duan, H., Sun, Y., Hu, R., Wu, B., Lin, J., ... & Zheng, H. (2022). Genome-wide association study with growth-related traits and secondary metabolite contents in red-and white-heart Chinese fir. *Frontiers in Plant Science, 13*, 922007.

Chanoca, A., et al. (2019). Lignin engineering in forest trees. *Frontiers in Plant Science, 10*, 912.

Chavan, S. B., et al. (2022). Enhancing farm income through boundary plantation of poplar (Populus deltoides): An economic analysis. *Sustainability, 14*, 8663.

Churkina, G. et al. (2020). Buildings as a global carbon sink. *Nat Sustain, 3*(4), 269–276

EU Commission. (2019). *Farm to fork strategy, for a fair, healthy and environmentally-friendly food system, Brussels* (pp. 16–19).

EU Commission. (2022). *Europäische Kommission: Neue EU-Waldstrategie für 2030.* Press release.

Dehoust, G., & Alwast, H. (2010). *Kapazitäten der energetischen Verwertung von Abfällen in Deutschland und ihre zukünftige Entwicklung in einer Kreislaufwirtschaft.* Öko-Institut.

Deutsche Bundesstiftung Umwelt. (2010). *Kurzumtriebsplantagen. Handlungsempfehlungen zur naturverträglichen Produktion von Energieholz aus dem Projekt NOVALIS.* https://www.dbu.de/phpTemplates/publikationen/pdf/120410114219pelp.pdf. Accessed on 18.08.2022.

Dev, I., et al. (2020). Bamboo-based agroforestry system (Dendrocalamus strictus+ sesame–chickpea) for enhancing productivity in semi-arid tropics of central India. *Agroforestry Systems, 94*, 1725–1739.

Dumke, H. (2020). *Erneuerbare Energien für Regionen: Flächenbedarfe und Flächenkonkurrenzen.* TU Wien Academic Press.

Ehling, R. (2020). *Strohballenbau. Vergleich dreier Bauvorhaben unter konstruktiven, ausführungstechnischen und bauphysikalischen Aspekten.* Master's thesis Leipzig University of Applied Sciences.

Espinoza, O., & Buehlmann, U. (2018). Cross-laminated timber in the USA: Opportunity for hardwoods? *Current Forestry Reports, 4*(1), 1–12.

EU. (2023). *Build in wood.* https://www.build-in-wood.eu/. Accessed on 30.9.2022.

Fachagentur für Nachwachsende Rohstoffe. (n. d). *Marktübersicht Hackschnitzelheizungen.* Accessed on 14.9.2022. http://www.fnr.de/fileadmin/allgemein/pdf/broschueren/Hackschnitzel-Heizungen_web_neu.pdf

Fachagentur für Nachwachsende Rohstoffe. (n. d.). *Energieholz.* Accessed on 14.9.2022. https://pflanzen.fnr.de/energiepflanzen/anbausysteme/energieholz. Last updated on 18.08.2022.

Fachagentur für Nachwachsende Rohstoffe. (2009). *Beschaffung.* https://nachhaltige-beschaffung.fnr.de/fileadmin/nawaro-kommunal/dateien/pdf_317-brosch_daemmstoffe2009.pdf. Accessed on 12.3.2023.

Fachagentur für Nachwachsende Rohstoffe. (2022a). *Zwei Fliegen mit einer Klappe: Pappelholz für Biomethan und Torfersatzstoffe.* Accessed on 1.11.2022. https://www.fnr.de/presse/pressemitteilungen/aktuelle-mitteilungen/aktuelle-nachricht/zwei-fliegen-mit-einer-klappe-pappelholz-fuer-biomethan-und-torfersatzstoffe

Fachagentur für Nachwachsende Rohstoffe. (2022b). Anbau nachwachsender Rohstoffe 2021 konstant. Accessed on 14.9.2022. https://www.fnr.de/presse/pressemitteilungen/aktuelle-mitteilungen/aktuelle-nachricht/anbau-nachwachsender-rohstoffe-2021-konstant. Last updated on 18.08.2022.

FAO. (2015). *Report of the regional meeting on agroecology in Sub-Saharan Africa. Dakar.*

FAO. (2020). *Global forest resource assessment, Rome.*

FAO. (n.d.). *Extent and characteristics of bamboo resources, Rome.* Accessed on 3.9.2022. https://www.fao.org/3/a1243e/a1243e03.pdf

Faust, K., & Schirmer, R. (2022). *Hybridpappeln: Mehr als nur KUP und Vorwald, Bayerische Landesanstalt für Wald und Forstwirtschaft, Munich.* Accessed on 31.10.2020. https://www.lwf.bayern.de/waldbau-bergwald/waldbau/248685/index.php.

Fernandez i Marti, A., & Dodd, R. S. (2018). Using CRISPR as a gene editing tool for validating adaptive gene function in tree landscape genomics. *Frontiers in Ecology and Evolution, 6*, 76.

Fladung, M., & Kersten, B. (2022). Tree genetic engineering, genome editing and genomics. *International Journal of Molecular Sciences, 23*, 22.

Fortier, J., et al. (2016). Potential for hybrid poplar riparian buffers to provide ecosystem services in three watersheds with contrasting agricultural land use. *Forests, 7*(2), 37.

LWF Freising. (2003). Der Energiegehalt von Holz und seine Bewertung. *LWF Merkblatt*, 12

Garnica. (2017). *Holzhackschnitzel, ein umweltfreundlicher Brennstoff*. Accessed on 19.8.2022, https://www.garnica.one/de/blog/holzhackschnitzel-ein-umweltfreundlicher-brennstoff.html. Last updated on 19.08.2022

HS Geisenheim. (2020). *Nachhaltiger Weinbau der Zukunft: Agro-Photovoltaik als weinbauliche Anpassungsstrategie*. Accessed on 3.9.2022. https://www.hs-geisenheim.de/forschung/institute/angewandte-oekologie/ueberblick-institut-fuer-angewandte-oekologie/forschung/

Göswein, V. et al. (2021). Land availability in Europe for a radical shift toward bio-based construction. *Sustainable Cities and Society, 70.*

Götzberger, A., & Zastrow. A. (1981). Kartoffeln unter dem Kollektor. *Sonnenenergie.*

Greene, C. H., & Scott-Buechler, C. M. (2022). Algal solutions: Transforming marine aquaculture from the bottom up for a sustainable future. *PLoS Biology, 20*, 10.

Grossarth, J. (2018). *Die Vergiftung der Erde, Metaphern und Symbole agrarpolitischer Diskurse seit Beginn der Industrialisierung, Frankfurt/New York* (pp. 159–166). Die kulturelle Wende der Agrarökologe, Darin

Grossarth, J. (2019). *Future food*. Die Zukunft der Welternährung, Darmstadt.

Grossarth, J. (2023). Agrarische Biomasse für den klimagerechten Bau, Ein Überblick über Nachhaltigkeitspotenziale Ressourcenpotenziale, Klimaschutzbeiträge, Anwendungen und Zielkonflikte. *nbau Nachhaltig bauen, 2*(2), 35–44.

Hartmann, H., et al. (2022). Climate change risks to global forest health: Emergence of unexpected events of elevated tree mortality worldwide. *Annual Review of Plant Biology, 73*, 673–702.

He, C. et al. (2024). Potentials of rice straw as feedstock for future bioeconomy: A case study from Hunan, China. Available at SSRN: https://ssrn.com/abstract=4828873 or https://doi.org/10.2139/ssrn.4828873

Herrmann, C., Idler, C., Heiermann, M. (2016). Biogas crops grown in energy crop rotations: Linking chemical composition and methane production characteristics. *Bioresource Technology, 206*, 23-35.

HLPE. (2019). *Agroecological and other innovative approaches for sustainable agriculture and food systems that enhance food security and nutrition, Rome.*

Fachwissen Holz. (2023). *Pappel*. Accessed on 30.11.2022. https://www.holzvomfach.de/fachwissen-holz/holz-abc/pappel/

Dach & Holzbau. (2019). *Holzstockwerk*. Accessed on 12.2.2023. https://www.dach-holzbau.de/artikel/18-stockwerke-aus-holz-3334141.html

Holzmann, G. et al. (2009). *Natürliche und pflanzliche Baustoffe*, Wiesbaden.

Ikromov, I., & Kadamov. A. (2021). Agroforestry against wind erosion damage: A case study in Tajikistan. In L. Müller (ed.), *Exploring and Optimising Agricultural Landscapes, Cham* (pp. 693–705).

Illampas, R., et al. (2014). Adobe bricks under compression: Experimental investigation and derivation of stress–strain equation. *Construction and Building Materials, 53*, 83–90.

Jang, H.-A. et al. (2021). CRISPR-knockout of CSE gene improves saccharification efficiency by reducing lignin content in hybrid poplar. *International Journal of Molecular Sciences,* 22, 18

Knaack, U. et al. (2022). *Bauen mit Papier, Architektur und Konstruktion, Basel.*

Kosiński, P., et al. (2022). Thermal properties of hemp shives used as insulation material in construction industry. *Energies, 15*, 7.

Kumar, D., & Mandal A. (2022). Review on manufacturing and fundamental aspects of laminated bamboo products for structural applications. *Construction and Building Materials, 348.*

Landgraf, D., et al. (2020). Biomass yield of 37 different SRC poplar varieties grown on a typical site in North Eastern Germany. *Forests, 11*, 10.

Lehmann, L. et al. (2020). Productivity and economic evaluation of agroforestry systems for sustainable production of food and non-food products. *Sustainability, 12*, 13

Lewandowski, I. et al. (1999). Miscanthus: European experience with a novel energy crop. *Biomass and Bioenergy, 19*, 4, S. 209–227.

LfL. (2022). *Press release.* Accessed on 12.1.2023. https://www.lfl.bayern.de/verschiedenes/presse/pms/2022/309222/index.php

LfL. (2023). *Grünland.* Accessed on 14.4.2023. https://www.lfl.bayern.de/ipz/gruenland/022269/

Li, Q., et al. (2022). CRP Involved in Nile Tilapia (Oreochromis niloticus) against bacterial infection. *Biology, 11*, 8.
Lignovis. (n.d.). *Versorgung mit Holzhackschnitzeln.* Accessed on 19.8.2022. https://lignovis.com/leistungen/versorgung-mit-holzhackschnitzeln.html
Lippe, M. et al. (2017). Die Herkunft der Biomasse. In Pietzsch, J. (Ed.), *Bioökonomie für Einsteiger*, Heidelberg/Berlin, pp. 11–66.
Lippe, M., Lewandowski, I., Unseld, R., Pucher, J., & Bräutigam, K. R. (2017). Die Herkunft der Biomasse. *Bioökonomie für Einsteiger*, 11-65.
van der Lugt, P. et al. (2018). *Carbon sequestration and carbon emissions reduction through bamboo forests and products.* International Bamboo and Rattan Organisation, Beijing
Lutter, R. et al. (2021). Climate benefit of different tree species on former agricultural land in northern Europe. *Forests, 12*
Lützkendorf, T., & Frischknecht, R. (2020). (Net-) zero-emission buildings: a typology of terms and definitions. *Buildings and Cities, 1*, 1.
Maalouf, C., et al. (2018). An energy and carbon footprint assessment upon the usage of hemp-lime concrete and recycled-PET façades for office facilities in France and Italy. *Journal of Cleaner Production, 170*, 1640–1653.
Mamanpush, S. H. et al. (2023). The impact of wood fibers in composite panels made from recycled fiberglass wind turbine blades. *Waste and Biomass Valorization*, pp. 1–8.
Marcar, N. E. (1987). Salt tolerance in the genus Lolium (ryegrass) during germination and growth. *Australian Journal of Agricultural Research, 38*(2), 297–307.
Marko, G., et al. (2016). Cross-laminated timber made of Hungarian raw materials. *IOP Conference Series in Materials Science Engineering, 123*, 12059.
Meyer, M., et al. (2021). Sustainable biomass value chains based on poplar plantations in European rural areas. *BioEnergy Research, 14*(2), 355–356.
Mohanty, A. K., et al. (2022). Sustainable polymers. *Nature Reviews Methods Primers, 2*(1), 46.
Moshelion, M. (2020). The dichotomy of yield and drought resistance: Translation challenges from basic research to crop adaptation to climate change. *EMBO Reports, 21*, 12.
Neugebauer, R. (Hrsg.) (2019). Biologische Transformation. Springer.
Nkeuwa, W. et al. (2022). Bamboo-based composites: A review on fundamentals and processes of bamboo bonding. *Engineering.*
Papiertechnische Stiftung PTS. (2015). *Faser und Papier*2030, n.p.
Pearson, R. G. (2006). Climate change and the migration capacity of species. *Trends in Ecology & Evolution, 21*(3), 111–113.
Pereira, H. M. et al. (2010). Scenarios for global biodiversity in the 21st century. *Science, 330*(6010), 1496–1501.
Pietzsch, J., & Schurr, U. (2017). Einführung. In This (Ed.), *Bioökonomie für Einsteiger*, pp. 1–10
Pillen, K. et al. (2020). Pflanzenbasierte Bioökonomie. In D. Thrän, and U. Moesenfechtel (Eds.), *Das System Bioökonomie*, pp. 35–49.
Poganietz, W.-R. et al. (2020). Zukünfte der bioökonomie: Eine modellbasierte analyse möglicher transformationspfade. In W. Konrad, (Ed.), *Bioökonomie nachhaltig gestalten: Perspektiven für ein zukunftsfähiges Wirtschaften.* Cham, pp. 187–222.
Pörtner, H.-O., et al. (2022). *Climate change 2022: Impacts, adaptation and vulnerability, IPCC 2022.*
PRS. (2023). *Sheffield.* Accessed on 22.2.2023. http://www.prsarchitects.com/projects/arts-civic/sheffield-winter-garden
Ratner, H. K. et al. (2016). Overview of CRISPR–Cas9 biology. *Cold Spring Harbor Protocols, 12*
Röck, M. et al. (2020). Embodied GHG emissions of buildings—The hidden challenge for effective climate change mitigation. *Applied Energy*, 258.
Rockström et al. (2016). The world's biggest gamble. *Earth's Future, 4*(10), 465-470.
Rüther, N. (2022). Naturfaserdämmstoffe, *Bauphysik Kalender 2022: Holzbau*, pp. 121–137.
San Miguel, G., et al. (2015). Environmental, energy and economic analysis of a biomass supply chain based on a poplar short rotation coppice in Spain. *Journal of Cleaner Production, 94*, 93–101.

Santolin, L. et al. (2022). Substrate-flexible two-stage fed-batch cultivations for the production of the PHA copolymer P (HB-co-HHx) with with cupriavidus necator Re2058/pCB113. In Mozejko-Ciesielska J. et al. (Eds), *Advances and Trends in Polyhydroxyalkanoate (PHA) Biopolymer Production* (2022), *Front Bioeng Biotechnol*, vol. 10.
Schäfer, D. (2021). *Wellerlehm*. In Schäfer, D. (Ed), *Massivbauweise mit Lehm*, Wiesbaden, pp. 47–57.
Schänzlin, J., & Grossarth, J. (2022). *Unpublished research proposal, working title: RePop, Institute for Timber Construction of Biberach University*
Schellnhuber, H.-J. (2021). Lecture given at the Zukunftsforum Bau. *Bonn, 18*(11), 21.
Schug, A.-L. (2022). *Schnellwachsendes Holz aus Bayern: Ein Beitrag zur Energiewende*. Bayerischer Rundfunk Online, 06.05.2022. https://www.br.de/nachrichten/bayern/schnellwachsendes-holz-aus-bayern-ein-beitrag-zur-energiewende,T4sgNiZ. Last checked on 19.08.2022
Solomon, T., et al. (2020). Promoting bamboo-based agroforestry for enhancing ecosystem services from degraded lands. *Agroforestry for Degraded Landscapes: Recent Advances and Emerging Challenges, 2*, 423–444.
Stolarski, M. et al. (2020). Energy value of yield and biomass quality of poplar grown in two consecutive 4-year harvest rotations in the North-East of Poland. *Energies, 13*, 6, 1495
Storaenso. (2021). *Technische Broschüre CLT*. Accessed on 12.6.2022. https://www.storaenso.com/-/media/documents/download-center/documents/product-brochures/wood-products/clt-by-stora-enso-technical-brochure-de.pdf
THessen. (2023). *Technologieland Hessen*. Accessed on 31.12.2022. https://www.technologieland-hessen.de/bp-biowert
Tol, R. (2002). Estimates of the damage costs of climate change, Part II. Dynamic estimates. *Environmental and resource Economics 21*, 2
Trommsdorff, M. et al. (2021). Combining food and energy production: Design of an agrivoltaic system applied in arable and vegetable farming in Germany. *Renewable and Sustainable Energy Reviews, 140*
Türk, O. (2014). *Stoffliche Nutzung nachwachsender Rohstoffe*. Springer.
UBA. (2022a). *Treibhausgasminderungsziele Deutschlands*, s.l.
UBA. (2022b). *Dauergrünland*. Accessed on 1.9.2022. https://www.umweltbundesamt.de/bild/gesamtflaeche-von-dauergruenland-anteil-an-der-0;
UBM. (2022). *Woho Berlin*. Accessed on 27.2.2023. https://www.ubm-development.com/magazin/woho-berlin/
Umweltbundesamt. (2010). *Globale Landflächen und Biomasse—Nachhaltig und ressourcenschonend nutzen*. Accessed on 18.8.2022. https://www.umweltbundesamt.de/sites/default/files/medien/479/publikationen/globale_landflaechen_biomasse_bf_klein.pdf. Last checked on 18.08.2022
Unnerstall, T. (2021). *Faktencheck Nachhaltigkeit*.
Vanhercke, T., & Colgrave, M. (2022). Future food production: What's brewing? Precision food proteins from fermentation. *Food Australia, 74*(1), 23–25.
Veste, M., & Böhm, C. (2018). *Agrarholz—Schnellwachsende Bäume in der Landwirtschaft. Biologie—Ökologie—Management*. Berlin/Heidelberg
Vieira, F. R., et al. (2022). Lignin as a renewable building block for sustainable polyurethanes. *Materials, 15*(17), 6182.
Wang, S. et al. (2020). Recent global decline of CO2 fertilization effects on vegetation photosynthesis. *Science, 370*(6522), 1295–1300.
Wei, P., et al. (2019). An exploratory study of composite cross-laminated timber (CCLT) made from bamboo and hemlock-fir mix. *BioResources, 14*(1), 2160–2170.
Wezel, A., et al. (2009). Agroecology as a science, a movement and a practice. A review. *Agronomy for Sustainable Development, 29*(4), 503–515.
Wong, J., et al. (2024). 3D printing with bamboo: An early-stage exploration towards its use in the built environment. *Sustainability, 16*(11), 4619.

Wu, M., et al. (2020). The moso bamboo drought-induced 19 protein PheDi19-8 functions oppositely to its interacting partner, PheCDPK22, to modulate drought stress tolerance. *Plant Science, 299*, 110605.
Xu, X. et al. (2021). Where and when are plantations established? Land-use replacement patterns of fast-growing plantations on agricultural land. *Biomass and Bioenergy, 144*, 105921
Yan, X. I. A. O., et al. (2022). Research progress of cross-laminated timber and bamboo (CLBT). *Journal of Building Structures, 43*(11), 126.
Yan, K., et al. (2024). Review on low-carbon development in Chinese industrial parks driven by bioeconomy strategies. *Renewable and Sustainable Energy Reviews, 199*, 114541.
Zhang, X., et al. (2022). The roadmap of bioeconomy in China. *Engineering Biology, 6*(4), 71–81.

5 Innovations in Bioeconomy: Mycelium Composites, 3D Printing, and Bio-Resins

Abstract

This chapter focuses on three specific examples of biotechnological innovations that are likely to be particularly relevant for the construction industry. Ultimately, they all revolve around the idea of breaking down wood and agricultural raw materials into very small components (whether to nanoparticle size or smaller), modifying them thermally, mechanically, or chemically, and thus creating new building material properties. Finally, the perspective shifts from technological to socio-economic value-chain innovations, whose establishment could enable the widespread adoption of sustainable innovations in practice.

The following discusses three specific examples of biotechnological innovations that are likely to be particularly relevant in practice for the construction industry. Ultimately, they all revolve around the idea of breaking down wood and agricultural raw materials into very small components (whether at nanoparticle size or below), modifying them thermally, mechanically or chemically and thus creating new building material properties.

5.1 Additive Manufacturing in Construction

The addition of traditional construction methods with solid wood construction is not only moving in the direction of increasing the use of glued wood materials such as glulam or cross-laminated timber (CLT). Materials based on lignin and cellulose, which are created based on changes in wood structures on a much smaller particle level, should also be considered. Perhaps such innovations are even the core of what should sensibly be understood as the bioeconomy in construction in the technical sense. The traditional construction and material wood is, so to speak, broken down into its smallest building blocks, or these are transformed in such

J. Grossarth, *Bioeconomy of Buildings*, https://doi.org/10.1007/978-3-031-84014-2_5

a way that completely new properties emerge. The raw materials for this can be (old) wood, grasses or agricultural biomaterials.

In the case of 3D printing (or the future technology of "4D printing"), cellulose is used as a (relatively inexpensive) filler material (Krapež Tomec and Kariž 2022). In 3D printing (or: "additive manufacturing"), lignocellulose serves as a functional additive and as reinforcement. On the one hand, the material properties of lignocellulose 3D printing need to be improved until they are ready for the market, on the other hand, the costs of production need to be significantly reduced (Lamm et al., 2020). However, they have already proven themselves to some extent on a laboratory scale.

Common 3D printing processes in the industry are based on carbon and glass fiber materials and fossil-based polymers. With concrete as a basic material, however, the first houses based on additively manufactured components have already been built—in the United States, in Belgium and Germany: In Beckum in Westphalia (House3D, 2020), a show house was completed in 2020—albeit still with 200,000 euros in government funding. For this, not biogenic raw material, but concrete ("I Tech® 3D from Heidelberg Cement) was used. As a significant advantage of such manufacturing processes the literature mentions.

- the high possibility of individualisation (from mass-produced prefabricated houses to unique pieces),
- less labour-intensive, mechanized production,
- on-site production,
- or the modular prefabrication.

Whether exterior walls like those of the house in Beckum can be printed on a biogenic basis and what the moisture and load-bearing behaviour of such houses will be is largely unknown. However, an application in the interior, for interior walls for example, or in furniture construction, seems likely. From the perspective of bioeconomic-circular thinking, these materials are interesting, because the innovative lignocellulose 3D printing could create structures that meet the requirements of circular, cascading material flow management (Kromoser et al., 2022). However, it must be asked on a case-by-case basis how exactly this claim is also fulfilled, or not.

In biogenic 3D printing, renewable raw materials can be used in different ways. A thermoplastic component as a cross-linker comes together with a cellulose material. Table 5.1 illustrates that the first material group can also be of biogenic origin:

The "bioplastic" PLA (Sect. 3.3) is an example of how polymers converted from waste fats can also be transformed into bio-based lignocellulosic raw material. The innovative bioeconomy thus bridges material boundaries between sectors in a historically unprecedented way. The important mechanical properties of the additively manufactured bio-based building materials (stiffness, strength, elasticity) can be altered by the recipes; cellulose can be used in the form of nanocrystals or nanofibers, and "filler contents of less than 30 weight percent of cellulose

Table 5.1 Overview of various 3-D printing processes. According to Lamm et al. (2020), Table 2.1

3-D printing process	Material 1 (thermoplastic)	Material 2 (lignocellulose)	Additives
Extrusion-based (layer-by-layer application of polymers) Production of meltable filaments	Thermoplastics (PLA, ABS, PP, Nylon, Polyurethane) Hydrogel material (Collagen, Carrageenan) or Nanocellulose in aqueous suspension	Wood flour; Spray or freeze-dried cellulose nanofibers; microcrystalline cellulose; lignin	Pigments, adhesion promoters, lubricants, glass, carbon
Granular Material Binding (FGF) (direct use of plastic granules instead of filaments) Selective laser sintering	Thermoplastics; cement-bound materials (Gypsum, Sodium Silicate, Portland cement)	Wood flour, wood chips, lignin;	Water
Water activation	Thermoplastic emulsion; resin	Sawdust, wood flour, methylcellulose	Water

nanofibers are most promising, with larger amounts leading to a decrease in all three mechanical properties" (Lamm et al., 2020).

The novel layer printing process *Individual Layer Fabrication* (ILF) allows the production of comparable materials with conventional chipboards, but with fewer binders required (Fig. 5.1). In this process, individually contoured wood-based panels are additively formed by the thin-layer application of wood particles (Henke et al., 2021). An advantage over other additive manufacturing processes is that mechanical pressure can be applied, thus achieving higher strengths of the material with less binder use. It is also possible to create unfilled cavities, which is particularly interesting for use in lightweight structures and the functionalisation of components (ibid.) The following process steps must be observed: First, wood chips are scattered, then locally provided with adhesive; the adhesive is hardened under mechanical pressure, then so bound with unbound wood chips are separated, finally, the 3D printer layers and laminates the material into 3D components (Bunze, 2023). A replacement of common, still used formaldehydes by protein- or starch-based binders is aimed for; a use of old wood or damaged wood should improve the ecological balance (ibid.; Aßhoff et al., 2021), also waste paper residues would be an option.

Fig. 5.1 An ILF sample element from the 3D printer. Photo by Jan Grossarth

5.2 Wood Functionalisation

Another branch of engineering research, which seems very promising from the perspective of the circular bioeconomy of buildings, deals with the creation of new material functions of wood building materials. This can concern the electrical conductivity, so that sensory light switches can be made from wood materials, for example (Dreimol et al., 2022), or it can concern the water permeability or waterproofness (Burgert, 2023). Changes in the lignin and cellulose contents are a key to this, and various composite materials with cellulose components are also being researched.

Nanocellulose is particularly in focus here as a "new material". The idea is to break down the wood into the smallest components and then reassemble them depending on the desired application area. The critical points here, however, are the recyclability, cascading ability, and the ecological balance of the resulting materials. The latter is significantly influenced by the energy expenditure of the heating or pressing processes including the drying of the lignocellulose.

The building material scientist Ingo Burgert from ETH Zurich spoke during a lecture in Rosenheim, impressively of comparable applications in nanomedicine—as early as 2015, an artificial ear had been printed. But an exterior wall is not an ear. In terms of possible uses, for example in the structure, he was more reserved—also towards 3D printing, because here he seemed to appreciate the properties of solid wood. He said: "The tree is our 3D printer, it makes us the structure" (Burgert, 2023).

Nevertheless, Burgert recognized a wealth of interesting research results in this: After the removal of lignin, cellulose becomes deformable, albeit with increased porosity, but with preserved fiber structure. From this, highly strong cellulose components for construction purposes can be formed by adding polymers and densifications through impregnations in resin (*densified cellulose materials;* Frey et al., 2018; Frey et al., 2019). Strength and stiffness can be varied. Highly water-repellent facade claddings could be an interesting application in construction. Wood materials with so-called *Janus membranes* make use of the directed water transport of the wood; with laser procedures, one side of the wood membrane is made hydrophobic, the other hydrophilic—one is therefore permeable to water, one is water-repellent (Ding et al., 2020). This allows breathability and moisture protection properties to be varied almost at will, compared to natural woods and allows applications perhaps even in civil engineering. By selective lignin removal and polymer additions, wood materials with improved insulation properties or altered thermal conductivity were also developed (Mi et al., 2020). In a way, wooden beams with thermal properties like "insulation bricks" can be produced in this way—possibly with improved recycling possibilities compared to styrofoam-filled insulation bricks. Such procedures can strengthen porous and light woods, and this with up to 80% slimming. This would benefit the ecological goal of conserving space. The densified wood also has significantly improved fire protection properties (Ding et al., 2020). Wood materials can also be made transparent like window glass; even coatings of photovoltaic cells have been developed (ibid.). In principle, all emission-intensive building materials could be functionally replaced by these methods (Fig. 5.2)—the question of raw material availability, energy expenditure and energy costs once excluded. If these methods reach market maturity, the choice of raw materials is likely to be guided by the lignin and cellulose contents of the wood species in addition to the prices (Table 5.2).

5.3 Bio-Based Adhesives

The reliance of innovative building materials made of wood—or more broadly: from lignocellulose—on synthetic adhesives, but also on the addition of flame retardants, antifreeze agents or fungicides, is repeatedly the main problem in terms of the desired recyclability and the possibilities for cascading downcycling. On the one hand, plug connections or screw and dowel connections, as they were already used in traditional timber construction, can be a solution—including unusual ones such as the plug connections tested in timber construction as "Japanese connections" (Discoue and Quinn 2022).

Biobased resins and adhesives, on the other hand, have further applications. They are likely to play a role especially in the context of the circular innovations mentioned above, although it would have to be clarified in individual cases whether the "renewable" is also the biochemically healthier means. For example, formaldehyde, a glue ingredient classified as carcinogenic by the EU and used in

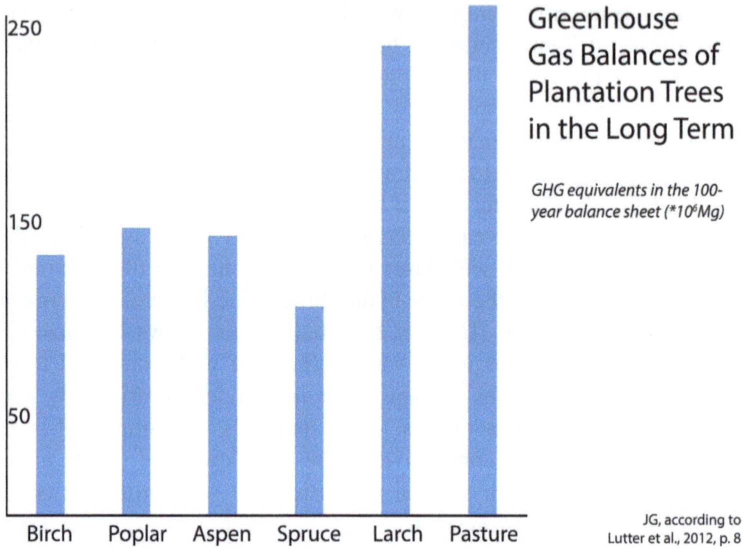

Fig. 5.2 Various innovations make it possible for steel, concrete or glass components to be replaced by wood-based building materials. JG after (Ding et al., 2020)

Table 5.2 Cellulose and lignin contents of trees. According to Türke (2014), p. 175

Tree species	Cellulose content (in %)	Lignin content (in %)
Pine (*pinus radiata*)	37.4	27.2
Beech (*Fagus sylvatica*)	39.4	24.8
Douglas fir (*Pseudotsuga menziesii*)	38.8	29.3
Eastern Hemlock (*Tsuga canadensis*)	37.7	30.5
Birch (*Betula verrucosa*)	41.0	22.0
Eucalyptus (*Eucalyptus globulus*)	51.3	21.9
Balsa (*Ochroma lagopus*)	47.7	21.5

large quantities in chipboard, also occurs "naturally". However, the approach of renewable raw materials is relevant in any case in terms of climate policy.

In this case, the construction industry is reverting to raw material sources for adhesives that were already used in pre-industrial centuries. As early as the Neolithic Age, *birch pitch* is said to have been used as an adhesive for attaching roof shingles made of bark, as seen in prehistoric buildings in Southern Germany. *Tree resins* were used as adhesives, as were *glutin* from boiled animal skins, glues from the cheese protein *casein*, natural asphalt from petroleum and others.

Synthetic resins only became established from the 1940s onwards and made bio-based adhesives obsolete. This was preceded by research successes: in 1909, the process of phenolic resin curing was patented, and in 1932, a ready-to-use adhesive based on this was introduced to the household market; the history of structural adhesives began somewhat later (Industrial Adhesives Association, pp. 1–19). Nature provides models. Proteins themselves have adhesive properties, a biological example being the protein-induced physical blood clotting (Rischka, 2007). With *Syndetikon* available from 1880 to at least the 1910s, a bio-based liquid adhesive (from the German Otto Ring & Co.) was mass-produced (Fig. 5.3). This was probably a vinegar-diluted, thickened glue (Hahn and Holfert 1906, p. 393), but according to other sources it was made from fish protein from boiled fish waste, sugar and glue (Braun, 2017, Chap. 1). It was on the German and international mass market. The exact recipe is not preserved, it was a trade secret. In the German Empire since 1897, under the brand name *Galalith* (literally meaning: "stone milk") based on casein as a protein and formaldehyde as a cross-linker, an early plastic based on a bio-based protein was marketed. This innovation is allegedly due to a cat that knocked over a bowl of formaldehyde in the inventor Adolf Spitteler's laboratory, which hardened in the cat's milk bowl. The raw material requirement for the production of this partially bio-based plastic was large: 6 percent of the milk produced in pre-War Germany or 30 million liters were used for this (Türk, 2014, pp. 130–132). However, these were not used for building materials, but for children's toys, combs or buttons.

Various new research works deal with the biogenic replacement of fossil-based adhesives. "Susbind" was from 2018–2022 the large European research project.[1] Among others, the Wood Competence Center of the research institution "Wood K Plus" in Linz and BOKU Vienna have been studying various adhesives based on biogenic raw materials for years, such as tannin from sweet chestnut, Radiata or Mimosa bark, lignin, plant protein, animal protein or plant oil (van Herwijnen et al., 2010; the same 2023). The long-term goal of such basic research is certainly also an industrial scalability—after all, it is about a significant replacement of large quantities of mostly fossil-based adhesives, such as the approximately 4 million tonnes used by the European chipboard industry alone of synthesised formaldehydes, which are produced on a petrochemical urea basis (ibid.). There are certainly (again) adhesives on the market that pick up old raw material traditions, and which are partially bio-based—like chipboards glued with soy-based adhesives (see box).

Lignin in particular is the focus of research due to its mass availability and structural similarity with phenol–formaldehyde resins (Averina et al., 2021). The field of application is broad and ranges from chipboard to asphalt binder. Proteins are also already being used as an adhesive for wood-based materials, mostly from waste products of starch or vegetable oil production. Carbohydrates such as starch and sugar are also being intensively researched as a raw material source, but there are hardly any marketable results yet (from Herwijnen, 2023). Press technologies play a key role (Solt et al., 2019). The scientific search for a resin synthesis on the basis of renewable raw materials was not yet completed in 2023. Lignin is suitable for road or street construction as a raw material for the addition as so-called bio-bitumen (Fig. 5.4).

Fig. 5.3 The early bio-based adhesive Syndetikon as advertising motif (undated). Photo Delcampe

Soy-based chipboards: The Purebond example.

In the United States, Columbia Forest Products is the first manufacturer to use a widely marketed soy protein adhesive ("PureBond") for plywood panels that are traded on the North American market; however, the high fluidity (viscosity) of the adhesive stands in the way of European application and many other applications in the construction industry. The plywood made from soy protein has comparable high waterresistance, shear strength, and durability compared to formaldehyde-based resins (Kennedy, 2017). The large American plywood manufacturer Columbia Forest Product markets

Fig. 5.4 Lignin-based bike path: A research path in Wageningen. Photo by Jan Grossarth

it nationwide. In Europe, it is not marketable because the EU standards for plywood impose stricter requirements, and the cross-linking agent used Kyemne® is classified as carcinogenic. Because "something that reacts well chemically, also reacts well with people, which means it is toxic," said Hendrikus van Herwijnen (2023) in his lecture. Also, Herwijnen said: "Every adhesive is an exploded pharmacy." Certainly, this applies not only to fossil-based, but also "natural" adhesives.

5.4 Case Study: Mycelium in Construction

Fungi are also suitable for biogenic material transformation. The varnished conk is considered an excellent biocatalyst. It is known from the central European and Chinese forests. The tree fungus prefers to settle on tired deciduous woods and decomposes them. The root networks, consisting of countless so-called hyphae, produce—when dried and hardened—a material that also contributes to house construction. The varnished conk or "Ganoderma lucidum", a rather slimy appearance in autumn, is also considered a remedy for neuroses in Chinese medicine. Within the framework of the European "Clean Industrial Deal", however, it is supposed to contribute to the path towards a circular bioeconomy. That is, to the fact that building materials can be reused better and with less pollution. And thus also, that construction emancipates itself from concrete.

In Karlsruhe, Professor Dirk Hebel for Sustainable Construction researches with varnished conk. "We want to grow a new class of building materials here that can replace metallic and mineral substances," explains Hebel (Grossarth, 2022). The architect at the Karlsruhe Institute for Technology (KIT) is considered one of the pioneers with this building material. He began his research on bamboo (Hebel et al., 2014) as a renewable building material, and a few years later discovered with others the varnished conk as a useful addition (Heisel et al., 2017).

But the hyphae of oyster mushrooms and many other types of fungi can also be used for material experiments. The fungi are given sawdust or other agricultural waste as a nutrient substrate, such as corn or straw. They must contain cellulose, hemicellulose, and lignin. The fungi then break down these substances into their components. They use enzymes for this. Each fungus ideally needs a special feed. This explains, in the forest, the occurrence of so many different forest mushrooms; they have specialized in different tree stocks. And each species has its own requirements for room temperature and humidity. Cultivation is not child's play even for scientific purposes: sterile conditions must always prevail in the laboratory, an encounter of the mycelium with a black mold would be fatal for future building materials, because no one wants to have these mold spores in their living room.

An advanced development from Hebel's team at KIT is the *New Wood*, which also received an award from the German Sustainable Building Council (DGNB). The cellulose and mushroom-based building material has similar properties to chipboard, but it does not contain toxic and difficult to compost adhesives. The New Wood consists of mushroom mycelium as a binder, which has grown on organic waste from agriculture, then dried and thus killed and hardened. Its production is, as they say: CO_2-negative. Its guiding principle: We should make waste materials usable for construction.

> We are now coming from years of basic research to a point where we have the first tangible products. My dream would be that we can choose in a few years in the hardware store between wood materials bound with synthetic adhesives or mushroom-based wood material panels.
>
> Dirk Hebel (2022)

A core problem is that the mushroom-based building materials are not very hard and often do not withstand even mild stress tests. To increase the tensile strength, they want to learn from concrete. It also has a low tensile strength and therefore needs reinforcement. For about a hundred years, it has therefore been poured onto steel reinforcements for house or bridge construction. The problem: After a later demolition, such a material mixture is only separable with great effort. At KIT, there are also research projects in which textile-based, organic reinforcements for the mushroom materials are used—such as hemp or cotton.

At RWTH Aachen, the civil engineer and professor for structural design Martin Trautz is researching with colleagues for the same reason on ways to reinforce the mycelium (Trautz et al., 2022). The mushroom-based composite materials of RWTH contain for the reinforcement of strength chips or larger pieces of wood, or laminated wood. Others contain pebbles or gravel. These too can be easily separated from the biomass. “I am convinced that we will one day be able to adjust the strength of the mycelium-based masonry bricks quite well,” predicts Trautz. However, they would still never be able to achieve the compressive strength of bricks or sand-lime bricks can. Until then “we continue to simmer in our witch’s kitchen,” he says—so further comprehensive series of tests, experiments with new mushroom cultures, and also first construction projects as a practical test are needed.

The scientists in Aachen are not lacking in imagination either. They talk about growing stones that connect themselves, for example. And the problem of lack of hardness can be reduced by layered and time-delayed growth of the cultures, writes Trautz. Here his building material tests resulted in a 175% increased load-bearing capacity. Also the use of wooden frames, into which the fluffy mushroom growth can be stuffed, seems to be a promising way. In addition, the choice of mushrooms can bring great improvements in strength: The tinder fungus produces up to 40% stronger materials, as it has a 20% higher chitin content, write Trautz and his colleagues in a current overview article in the journal “Nachhaltig Bauen” (Sustainable construction). Chitin is the substance from which crab shells are made. The fact that mushrooms contain it makes them so interesting as a building material.

According to the scientists, infills with stones made of such mycelium composites, as they are researched in Aachen, should be feasible in the near future. But before they really arrive in the practice of building a single-family house, many official building material approvals and German or European standardisations must take place.

In addition, building material companies or others in this field must invest a lot of money: in large-scale plants, in research and development. And the end customers must also demand the mushroom stones. Many will want to wait and see what experiences others have with this organic building material, for example in terms of durability and moisture tolerance.

For mushroom-based insulation materials, however, the time should be ripe sooner—for example also in foam form. “We want to replace PU foams as the very first thing” says Professor Trautz from Aachen. He is referring to polyurethane foams, which have been used for decades as sealants, insulation materials, mortar substitutes, or adhesives for polystyrene insulation boards in construction. These not only pose a potential health risk (Werner, 2020), but also a problem when it comes to separating and recycling the components after demolition.

Dirk Hebel in Karlsruhe also refers to foams as "low hanging fruits" that will soon be harvested. His dream house is compostable, and he imagines the house of the future to be light, multi-layered, and without purely mineral walls. Hebel speaks—following the model of half-timbered houses—of solid wood structures, filled in, with an insulation layer perhaps based on fungi, clay plaster as a heat storage layer. He also sees systems from renewable and biodegradable sources as future adhesives: "There has been no breakthrough here yet," says Hebel.

Solid wood will not be replaceable. "Without wood, it's not possible if you want load-bearing materials based on fungi," Hebel states today with regard to the desired strength values. His team is currently testing so-called wood veneer-reinforced mycelium composites, such as mixtures of varnished conk, maple veneer wood, and hemp. Fire resistance, moisture tolerance, or insulation values are also research topics. "We can bring the U-value to very good values, for example, through air inclusions," says Hebel.

Hebel's specialty remains reinforcements with bamboo fibers, which are then coated by fungi. He had already erected the first self-supporting structure from this, much noticed, five years ago. This was the time of the first experiments with the fungus as a building material. As early as 2014, the so-called Hy-Fi Tower was built in New York from corn waste and mycelia. And in 2019, Dutch architects and material designers erected another pavilion building in Eindhoven, reinforced with cotton fibers, the "Growing Pavilion".

The question of whether the whole project will ultimately be worthwhile is not only decided by material properties and prices of the cultivated building materials. An important question is also what the fungi "eat". Today, chips of spruce, beech, birch are often given. But another step towards a circular economy without waste is the possibility of also using old wood from the construction industry as food for the fungi. Last but not least, the question of efficiency is crucial. How fast do the fungi grow? How good are the material properties? Martin Ostermann, Professor of Building Construction at the University of Stuttgart, is working on this. His project Mycomodul, funded by the Bundesministerium für Bildung und Forschung (BMBF/eng.: Federal Ministry of Research), is about the use of mycelium in lightweight construction. The Stuttgart laboratory "Future Material Lab" looks like a large bakery, but more sterile. The processes: The mycelium composite is poured into a mold, where it grows, and is then finally baked and dried in the oven. "We are also researching its application in 3D printing," says Ostermann. Stuttgart specialties currently also include the textile formwork with which the fungi grow together. "Then you don't have to remove it, as with concrete construction, after pouring," explains Ostermann. He works with biopolymers or hemp for this.

The University of Stuttgart has a reputation for its research on so-called gradient components, such as lightweight concrete slabs with air bubbles. This saves material. Ostermann also wants to produce such gradient components based on fungi. As far as possible strength values are concerned, he is not very reserved: "We want to improve the filaments of the fungus, improve the quality of the fungus," he says—unlike American or Chinese scientists, without genetic engineering. "Because the material is supposed to be compostable, i.e. released back into nature." Therefore, the rule here is: Everything fully compostable. This is how our construction waste of the next generation is to become flower fertiliser.

Fungi for agriculture—Mycorrhiza as a way into an industrialized sustainable agriculture?

Fungal root networks, known as mycorrhiza, can be used in agriculture to increase soil fertility. They have been present in every natural soil for about 450 million years but have been greatly reduced in fields due to intensive farming over the past decades. Every plowing tears up the underground fungal roots. When monocultures are grown, as in the case of corn, the fungal world under the ground becomes impoverished, as many of the roots live in symbiosis with plants. If the fungi are missing, farmers have to bring more fertiliser to the field so that the plants can still grow. As has been known for several decades, these networks are something like suppliers for the plant; they add nutrients such as phosphorus and nitrogen to the roots, which they cannot reach on their own. Both the fungus and the grain benefit from this. It gains access to reservoirs of water and phosphorus that it could not reach with its roots alone. Because soils form little new humus, the number of fungi in the soil decreases, and monotonous crop rotations like in the corn growing areas for the biogas plants do not contribute to revitalisation (Schüßler et al., 2001; Redecker et al., 2013).

Fungi in symbiosis with the root and certain soil bacteria improve nutrient uptake and plant growth. If they are added to the soil, the balance of nitrogen and phosphorus fertilisation can be improved. Phosphorus is finite, estimates for the end of the deposits range from a few decades (TU Sydney) to about 300 years (US Geological Survey). Various industrial manufacturers have been offering mycorrhiza powder since the mid-2010s.

5.5 Case Study: Social and Value-Chain Innovations

A narrow interpretation of the concept of innovation in terms of technical aspects would be too short-sighted. New forms of cooperation and communication that lead to higher value creation are also to be described as (social or value chain) innovations. Such innovations are the basis for the hope of a "construction turnaround", but also for small farmers or development policy with Africa. Experiences from concrete projects show how successful cooperation along value chains can succeed.

While they still focus on the food sector, they may be transferable to the construction sector—including the one in Europe. But what is, first of all, to be understood by *innovation*? Is technical progress always innovation? No. The simply new is rarely innovative. This is the case with technology. But it is also the case with the living. A new child, for example, is not an innovation (even if each of the eight billion people on earth is unique). A new corn plant is not necessarily innovative (even if it was crossed from two different parent varieties and has new "use values", such as size and starch content). In the United States, corn yields have indeed increased significantly—but the number of hungry people there has not decreased as a result. As long as this remains the case, corn breeding—despite its historical contributions—is unlikely to be referred to as innovation in relation to the current hunger problem in the US. So, what is innovation? A technical innovation is innovative when it meets needs. This is the decisive ethical extension. Social innovations have a content dimension, a process or relationship dimension, an empowerment dimension (participation, education, access to resources). Perhaps it was a deficit of "western" thinking about innovation that they were viewed purely technologically (Mavhunga 2017). The word innovation has a long and revealing history. The linguist Susanne Weber has researched and mercilessly dissected the history of the term innovation in her 2018 Berlin dissertation. She sees it as a "stuffed goose term", a "flag word" or "magic potion". The claims of originality are rarely actually fulfilled. The word means in the political-ministerial language use often not much more than an acceleration of the new. And as such, according to Susanne Weber, it is a "significant reference point for government action". A comparable prime example is the often substanceless talk of digitalisation (Weber 2018). Also interesting is the historical change from predominantly negative to positive intention of word usage. From antiquity through the Christian Middle Ages to modern times, the word has been almost exclusively negatively connoted: in the sense of upheaval, revolution, heresy. "Nihil innovetur nisi quod traditum est," is a quote from early Christian times attributed to Pope Stephen I. "Nothing is innovative that does not build on tradition," one could translate somewhat freely.

But the history of the term also allows a connection to its bright sides. Development policy has increasingly remembered this in recent years: The quality of innovation is already hinted at in Schumpeter's "creative". No less interesting is a reference to an ancient dictionary from 1564, in which the Middle High German "Erneuwern", close to high-german "renewing", is listed as a synonym for improving, refreshing, reshaping, or "bringing back into the old form". From Joseph Schumpeter's (1911) differentiated economic-sociological analysis and his concept of "creative destruction" to banal ideals of "entrepreneurship", the term carries an increasingly one-sided focus on economic valorisation. Its use as a filler word brings scientists and politicians low-requirement reputation gains.

Innovations are therefore need-based novelties. The only question is: what exactly is a need? This is where ethics comes into play. A need can be material or immaterial, but not maliciously justified. Could one seriously say, "I had a need to kill my husband?" Or even, "I have a need to become rich." Not seriously. Positive values must be touched, from the right to life in dignity to creative development. This applies to all the UN Sustainable Development Goals (SDG). However, innovations do not exclusively bring progress for the good. They have an ambivalent effect: Almost every progress has its price. The invention of the automobile brought more traffic deaths and environmental pollution, but also new experiences through travel. Thanks to ambulances, there is faster help for the sick and injured (including those injured in car accidents). A binary coding of "need-based" or "need-unjust" is impossible. On the whole, a balancing of needs is required. This insight was the reason for the boom in ethics as a companion science to technology sciences from the 1980s onwards.

For decades, the idea of "Western" development aid was also based primarily on the export of techniques to the southern world. Whether it's seeds, machines, or institutions—"democratic", "socialist": Every seed can only sprout if the soil is fertile for it. Regardless of whether it's about tractors, genetic engineering, modern government structures, or new transport infrastructure: Every technical or institutional innovation is only an "innovation" where it contributes to solving local problems. Otherwise, all this ends in rusty machine graveyards, hollow and inflated administrative structures, dilapidated roads that were once built, but for whose maintenance the financial and fiscal foundations are lacking. Africa is full of examples: remnants of a past era of excessive innovation.

"Grow Africa" in South Africa, for example, is an initiative that can be described as a value-chain innovation. According to its own statements, Grow Africa has already accompanied more than a hundred companies in several African states. And in this way, using agriculture as an example: Companies and stakeholders define the so-called "cash crops" for a region—cassava, for example, or corn, peanuts, sweet potatoes. The initiative then focuses on these. Now all relevant actors come to the table: farmers' organisations and trade, logistics and processing, exporters, Politics, the so-called civil society. From the "field to the plate", products are created that promise higher value creation, and thus also provide sufficient income for farmers. To this end, the initiative brings investors from home and abroad into the chain. In order for them to truly believe in the success of the project, and for the individual members of the value chain to adhere to the agreements made, the group appoints a "Chief Investment Officer". The "Cassanovas" from Nigeria, chips made from cassava, have market successes just in Nigeria. Perhaps there are also certain export prospects in limited buyer milieus of the global North. But even the urban consumers in Nigeria's neighboring countries prefer

Fig. 5.5 How can value be generated from peasant products? Corn offer in Zambia. Photo by Jan Grossarth

their respective local products, cassava from Cameroon for example. And the brand called "It's Wild" in Zambia (Fig. 5.5) offers 17 products, for which thousands of small farmers provide the raw materials—the logo is a Zambian elephant. Because: "The most inspiring and urgently needed innovations come from people who respect, thoroughly understand and build on local forms of knowledge." (Mavhunga 2017).

Notes
https://susbind.eu/.

References

Acatech. (2020). *Holzbasierte Bioökonomie. Nachhaltig, zirkulär, klimaresilient.*

AltholzVo. (2020). *AltholzVo.* https://www.gesetze-im-internet.de/altholzv/AltholzV.pdf

Aßhoff, C., et al. (2021). Individual Layer Fabrication (ILF)—Additive Fertigung durch selektives Binden von Holzspänen zu Einzelschichten. *Holztechnologie, 62*(4), 5–13.

Averina, E. et al. (2021). Protein adhesives: Alkaline hydrolysis of different crop proteins as modification for improved wood bonding performance. *Industrial Crops and Products, 161.*

Braun, D. (2017). *Kleine Geschichte der Kunststoffe.* Hanser.

Bunze, F. (2023). Individual Layer Fabrication (ILF)—Ein neues Verfahren zur additiven Fertigung von Bauteilen aus Hol. In *Lecture given at the 1st International Wood BioEconomy Forum (WBE).* Rosenheim University of Applied Sciences.

Burgert, I. (2023). Funktionelle holzbasierte Materialien für das Bauwesen. In *lecture given at the 1st International Wood BioEconomy Forum (WBE), February 16, 2023*, Rosenheim University of Applied Sciences.
Coenen, J. et al. (2022). Expediting the implementation of closed-loop supply chain management: a facilitated case study on re-using timber in construction projects. *Circular Economy and Sustainability*, 1–32.
Ding, Y., et al. (2020). Janus wood membranes for autonomous water transport and fog collection. *Journal of Materials Chemistry, 8*(42), 22001–22008.
Discoe, P., & Quinn, A. (2022). *Zen architecture: The building process as practice, Utah.*
Dreimol, C. H., et al. (2022). Sustainable wood electronics by iron-catalyzed laser-induced graphitization for large-scale applications. *Nature Communications, 13*, 1.
Frey, M., et al. (2018). Delignified and densified cellulose bulk materials with excellent tensile properties for sustainable engineering. *10*(5), 5030–5037.
Frey, M., et al. (2019). Delignified wood–polymer interpenetrating composites exceeding the rule of mixtures. *ACS Applied Materials & Interfaces, 11*(38), 35305–35311.
Grossarth, J. (2022). *Ein Baustoff steht im Walde.* Welt am Sonntag.
Hahn, E., & Holfert, J. (1906). *Spezialitäten und Geheimmittel, Ihre Herkunft und Zusammensetzung. Eine Sammlung von Analysen und Gutachten*, Heidelberg.
Hebel, D. E. (2022). *Authorized quote from a telephone conversation by Jan Grossarth for an interview with "Welt am Sonntag", transcript.*
Hebel, D. E., et al. (2014). Process-controlled optimisation of the tensile strength of bamboo fiber composites for structural applications. *Composites Part B: Engineering, 67*, 125–131.
Heisel, F. et al. (2017). Design of a load-bearing mycelium structure through informed structural engineering. In *Proceedings of the World Congress on Sustainable Technologies (WCST).*
Henke, K., et al. (2021). Individual layer fabrication (ILF): A novel approach to additive manufacturing by the use of wood. *European Journal of Wood and Wood Products, 79*, 745–748.
House3D. (2020). *Houseprint.* Accessed on March 30, 2023. https://www.houseprint.de/
Industrieverband Klebstoffe e. V. (2017, Ed.). Ein langer Klebfaden durch die Vergangenheit. In: *Auf der Höhe der Zeit.*
Kennedy, E. B. (2017). Biomimicry. Design by analogy to biology. *Research-Technology Management, 60*(6), 51–56.
Krapež Tomec, D., & Kariž, M. (2022). Use of wood in additive manufacturing: Review and future prospects. *Polymers, 14*, 6.
Kromoser, B. et al. (2022). Circular economy in wood construction–Additive manufacturing of fully recyclable walls made from renewables: Proof of concept and preliminary data. *Construction and Building Materials, 344*
Lamm, M. E., et al. (2020). Material extrusion additive manufacturing of wood and lignocellulosic filled composites. *Polymers, 12*, 9.
Mavhunga, C. C. (2017). Introduction: What Do Science, Technology, and Innovation Mean From Africa?, *MIT Press,* 1–27.
Mi, R., et al. (2020). Scalable aesthetic transparent wood for energy efficient buildings. *Nature Communications, 11*, 1.
Miletzky, F. et al. (2020). Holzbasierte Bioökonomie. In D. Thrän & U. Moesenfechtel (Eds.), *Das System Bioökonomie* (pp. 51–68).
Redecker, D., et al. (2013). An evidence-based consensus for the classification of arbuscular mycorrhizal fungi (Glomeromycota). *Mycorrhiza, 23*, 515–531.
Rischka, K. (2007). *Proteine-Ganz gewöhnliche Klebstoffe, conference contribution with online abstract.* Accessed on November 22, 2022. https://publica.fraunhofer.de/entities/publication/9dd8f251-f092-4c68-a682-ac6140a16988/details
Risse, M., & Richter, K. (2018). *CaReWood–cascading recovered wood, teilvorhaben: Ökologische und ökonomische Bewertung der kaskadischen Holznutzung, Munich.*
Schüßler, A., et al. (2001). A new fungal phylum, the Glomeromycota: Phylogeny and evolution. *Mycological Research, 105*(12), 1413–1421.

Solt, P. et al. (2019), Technological performance of formaldehyde-free adhesive alternatives for particleboard industry, *International Journal of Adhesion and Adhesives*, 94, 99–131.
Teischiner, A. (2023). Spannungsfeld Altholz und Frischholzvorräte für Energie und Produkte? In *Lecture given at the 1st international Wood BioEconomy Forum (WBE), February 16, 2023.* Technical University of Rosenheim.
Teischinger, A., & Tiefenthaler, B. (2009). *Zukunftsideen für Forst-Holz-Papier, final report of the project.* Technologie-Roadmap für Holz in Österreich.
Trautz, M., et al. (2022). Myzelkomposit als vielseitig einsetzbarer, kreislaufgerechter Werkstoff für das Bauen. *Nbau. Nachhaltig Bauen, 1*(2), 37–45.
Türk, O. (2014). Stoffliche Nutzung nachwachsender Rohstoffe. Springer.
UBA. (2020). *Evaluierung AltholzV*. Accessed on January 9, 2023. https://www.umweltbundesamt.de/sites/default/files/medien/479/publikationen/texte_95-2020_evaluierung_der_altholzverordnung_im_hinblick_auf_eine_notwendige_novellierung.pdf
van Herwijnen, H. W. (2023). *Überblick neue Klebstofftechnologien.* Lecture given at the Holz-Bioökonomieforum.
Van Herwijnen, H. W., et al. (2010). Verwendung nachwachsender Rohstoffe in Bindemitteln für Holzwerkstoffe. *Chemie Ingenieur Technik, 82*(8), 1161–1168.
Wasteresearch. (2019). *Der Markt für Altholz in Deutschland bis 2030*, 2, n.p.
Weber, S. (2018). *Innovation: Begriffsgeschichte eines modernen Fahnenworts.* Tectum.
Werner, J. (2020). *Montageschaum—Gefahren und sichere Verwendung, Consumer portal of the Bavarian State Ministry for the environment and consumer protection.* https://www.vis.bayern.de/produkte_energie/basisinformationen/pu_schaeume.htm

Circular Economy in Construction: Current Status, Potentials, and Strategies

6

Abstract

The extraction and processing of mineral, metallic, and biogenic resources significantly contribute to global greenhouse gas emissions. Given ongoing urbanization and global increases in wealth, the total amount of raw materials extracted worldwide is expected to nearly double by the middle of this century. Resource conservation at all levels of the economy is essential. The construction industry is a central focus, as it generates the largest amounts of waste both in Germany and globally. Circularity conserves resources, and a multitude of measures can promote more careful resource management. These range from certifications for urban mining to establishing digital construction material exchanges and stricter waste regulations, such as take-back obligations for building materials or mandatory building material passports. Also relevant are adapted and new standards aimed at a circular economy, as well as changes in construction and planning processes and circular product design. The concept of *Cradle to Cradle* enhances the goal of continuous circularity of technical components with the ideal of a biosphere designed entirely for "compostability"—such as materials made from wood fibers or straw. The vision of a "world without waste" inspired by nature supports the goal of resource conservation but is ultimately ahistorical and impractical as a rigid norm. However, it is highly attractive to innovative companies and is an important part of the circular transformation. Another important factor is innovations in the waste and recycling industry, which must work with waste streams as they currently exist. One example is the sensory separation of construction material fractions based on color criteria from construction debris, which facilitates sorting and recycling. This article first discusses resource scarcity forecasts for building materials, recycling rates in construction, and provides an overview of resource-saving materials such as lightweight and gradient concrete, Celitement, or wood hybrid ceilings. It then examines the historical contexts of these concepts to clarify the differences between circular economy and the Circular Economy model. The article provides an introductory overview of the role of component exchanges

J. Grossarth, *Bioeconomy of Buildings*, https://doi.org/10.1007/978-3-031-84014-2_6

and material registries, take-back obligations for building materials, the potential of urban mining, and introduces the *Urban Mining Index* planning tool. It concludes with a look at prominent examples of architecture and a historical review of the careful deconstruction practices of the 1920s, which can serve as a model for a circular construction industry.

6.1 Summary

The extraction and processing of mineral, metallic and biogenic resources is a major contributor to global greenhouse gas emissions. The total amount of raw materials extracted globally is expected to almost double by the middle of the century. Resource conservation at all levels of the economy is necessary. The construction industry is in the central focus, as it generates the largest amounts of waste in Germany and worldwide. A circular economy conserves resources. A multitude of measures can lead to a more careful use of raw materials. They range from certifications for "urban mining" to the establishment of digital building material exchanges to stricter waste law requirements that, for example, establish take-back obligations for building materials or mandatory building material passports. In this context, it is also necessary to consider standards adapted to the circular economy as well as new standards up to changed construction and planning processes and a circular product design. The concept of *Cradle to Cradle®* complements the claim of permanent recycling of technical components with the ideal of a biosphere consistently oriented towards "compostability"—this includes, for example, building materials made of wood fibers or straw. The idea of a "world without waste" modelled on nature contributes to the orientation towards the goal of resource conservation but is ultimately ahistorical and impractical as a rigid norm.

6.2 Raw Materials and Reserves

> It is indeed only fragmentarily known which building materials will be needed where in the world in the next 20 years—and even less where they are available" Werner Sobek (2023).

6.2.1 Scarcity of Raw Materials

Construction is resource-intensive (Fig. 6.1) and the demand for the three largest bulk raw material groups is likely to continue to rise strongly globally in the coming decades (Fig. 6.2). A regional recycling of materials can contribute to resource conservation. Raw material exploitation and processing alone account for about 50% of global greenhouse gas emissions, and even more than 90% of biodiversity loss (UN EN 2019). Included in this calculation—besides metals, non-metallic

Fig. 6.1 Resource-intensive construction in the Frankfurt banking district. Photo Jan Grossarth

minerals such as stones and sand, and fossil fuels such as coal and oil—are also the agricultural and forestry raw materials.

The cultivation and extraction of this so-called biomass alone account for 17% of GHG emissions, and for the remaining raw material classes:

- metals account for 10%
- "fossils" account for 16% and
- non-metallic minerals account for 10% (ibid.).

Historical False Projections of Collapse

More relevant than the resources are, with a view to scarcities, the *reserves*, i.e. the economically recoverable and actually proven resources. The worries about resource scarcity are a phenomenon that can be traced back centuries and often proved deceptive in retrospect (Schmidt, 2019). Forecasts of their ranges have repeatedly failed historically, among other things because technical advances such as hot steam injection methods for oil production were not foreseeable. One of the reasons was that new deposits were found, or that increased resource efficiency reduced usage pressure. For example, the "Club of Rome" in its iconic report "Limits of Growth" 1972 famously still predicted that the global natural gas reserves would be exhausted by 2010, and those of oil as early as 2003. Both were wrong. Instead of scarcity scenarios, the reservation of GHG-emission-low extraction now occurs in many cases.

Fig. 6.2 Where will the raw materials run out first? Copper mine in the north of Zambia. And clay mining in the Hessian Spessart. Sketch shows global distribution of the "urban mine" for copper. Photos by Jan Grossarth; data https://outworldindata.org/

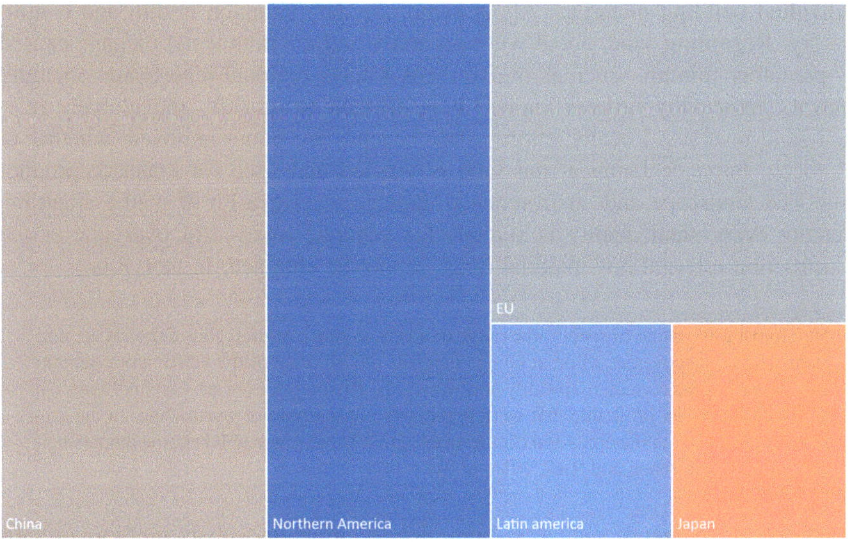

Fig. 6.2 (continued)

Raw Material Deposits by Countries

The German Federal Institute for Geosciences and Natural Resources has, according to various criteria, created a total index of the richest countries in raw materials globally. The top ten are (BGR, 2020, p. 18).

1. China
2. Brazil
3. Australia
4. Russian Federation
5. United States
6. Canada
7. India
8. South Africa
9. Chile
10. Indonesia
11. …
12. Germany.

6.2.2 Which Raw Materials Are or Will Become Scarce?

Is a blanket talk of resource shortages at all appropriate? With a view to the most important group of mineral building materials, this cannot generally be said. By far not all construction-relevant resources are scarce. Scarcity scenarios concern individual building materials. About some, the data situation is thin and contradictory: Regarding sand, about whose scarcity and environmental damage caused by predatory mining—such as on riverbeds—many regional reports are generally known, statistically little is known. It is difficult to quantify the globally relevant sand reserves. Locally, entire sandy beaches are being removed, whether in Morocco, India or Jamaica; the sand is washed and used for concrete production. The landscape and environmental damage receive a lot of media attention. But not every sand quality is suitable for concrete mixes. An overview of the construction-relevant raw material range cannot be obtained. In fact, it is

> but almost impossible to specify the range of a certain raw material. New deposits are constantly being discovered, while at known deposits the estimate of the existing ore quantity must regularly be corrected upwards or downwards. Rising prices or new technologies can also make deposits profitable that were previously considered not worthwhile. In the long term, deposits in remote areas and occurrences with low ore grade will become increasingly important (Neukirchen and Ries, 2016, p. 9).

Many critical metals such as indium, fluorite, graphite, cobalt, niobium or antimony are relevant for electronic applications, for example, of the minerals directly significant for the construction industry, according to the USGS the following static reserves are currently assumed:

- Graphite (about pulverized as an additive to increase the thermal conductivity): 291 years,
- Iron: 55 years,
- Copper: 43 years,
- Gypsum: 15 years.

The bulk building material gypsum will also become scarce and expensive in Germany because the phase-out of coal will deplete the gypsum board raw material FGD gypsum from the flue gas desulfurisation plants of coal-fired power plants (BR, 2022). From the perspective of the construction industry, the bulk raw materials are of particular interest (Neukirchen and Ries, 2016, pp. 317–332). More precise than global values, in the case of bulk construction materials, upcoming *regional* shortages can be predicted. For a climate-friendly construction, due to the transport emissions in many cases, the supply of regional raw materials is significant. If you divide the reserve of a raw material by the annual consumption, you get the *static range*. For the state of Baden-Württemberg for example, the range of reserves for relevant regionally extractable building materials in the case of a southern German region is as Table 6.1 shows:

Minerals or metals (Fig. 6.2) are also among the *raw materials with the highest supply risk for the German economy*. For *iron*, raw material reports predict a critical supply perspective (VBW, 2020, p. 14); but this list is also led by other substances such as rare earths: cobalt, tantalum, yttrium, rhodium, gallium, tungsten, tin, indium, niobium and others. Among the mineral building materials, only *gypsum* and *anhydrite* (as calcium sulfate, for example as additive for screeds) and

Table 6.1 Static range of building materials in Baden-Württemberg (LGRB 2019)

Cement raw materials (incl. oil shale)	50 years
Natural stone	50 years
Brick raw materials	< 50 years
High-purity limestones	< 30 years
Carbonate rocks	20 years
Sandy gravels	15 years

mica—a ceramic base material—are assessed as moderately critical (VBW, 2020, p. 101f.).

6.2.3 Construction Waste and Recycling (Germany as a Case)

The amounts of demolition material have been increasing in Germany for decades (Figs. 6.3, 6.4 and 6.5). This trend is likely to continue. Recycling or reuse of building materials is also necessary in view of the pure material fills at the demolition. About 200 of 340 million tons of German waste were construction and demolition waste (2013; according to Pietzsch 2017, p. 55). The construction industry is by far the largest source of waste by weight. At the same time, high recycling rates can be observed here. For construction and demolition waste this is 87%—in comparison: only 68% of municipal waste is recycled (Eurostat, 2021). A major reason for the comparatively high values is the circular economy legislation. Table 6.2 shows the recycling rates in construction.

It is noticeable that the very high recycling rate of 93.7% in the road construction sector improves the overall balance of the construction industry. In particular, the low recycling rates in the area of gypsum mining and soil excavation are striking. As far as soil excavation is concerned, a use as a component for clay bricks would be absolutely appropriate. Initial research projects exist. Clay bricks are considered circular or closed-loop material. They can be completely remade into clay bricks at the end of a building life cycle. A standardisation project for load-bearing clay brick masonry was completed in 2023 (DIN 18,940). There is a great potential to use the largest proportion of construction waste in parts in a high-quality way. The gypsum components in the construction rubble bring the problem of leading to unwanted loads with sulfates.

Construction rubble contains gypsum, and there has been an increasing and diverse use of gypsum in construction over the past decades. This is an obstacle to higher recycling rates. Separation techniques or pure construction could solve the problem—the former in the short and medium term, the latter in the long term, as the buildings built today are likely to be demolished in about 50–70 years.

The processing of construction rubble would first have to be able to separate more precisely, which is indeed more and more mixed or "glued". The construction practice has been tending towards ever thinner solid wall structures for decades, combined with numerous glued and dowelled insulation and facade protection

Fig. 6.3 Construction waste at a recycling plant in Southern Germany. Photo by Jan Grossarth

layers (thermal insulation composite systems). Processing of construction rubble begins with crushing, i.e. mechanical breakdown, followed by classification, sorting, cleaning, separation, crushing.

6.2.4 Reusability of Building Materials

Recycling and the reuse of raw materials are seen as central strategies in dealing with growing scarcities. In a publication on resource efficiency as a contribution to

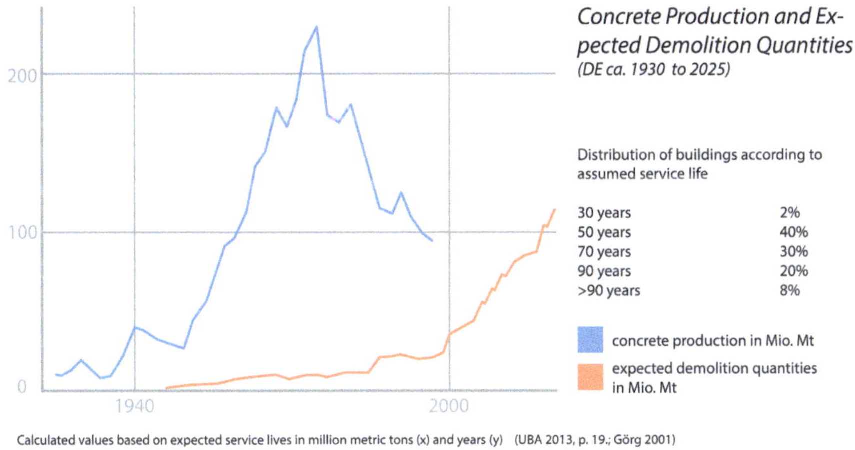

Fig. 6.4 Concrete rubble as building material of the future? A forecast of accumulating masses

Fig. 6.5 A treasure trove? Mixed, crushed construction waste. Photo Jan Grossarth

tackling the climate crisis, the United Nations Environment Programme (UNEP) lists as goal-oriented strategies for increasing material efficiency:

- Lightweight construction,
- Substitution of mineral building materials with wood,

Table 6.2 Fractions of construction and demolition waste in Germany (2014), according to BBS, Mettke et al., p. 127; * under "Other recovery" according to §6 KrWG the Law on Circular Economy, energy recovery, mine backfilling or filling of pits, quarries, open-cast mines is to be understood—only after prevention, preparation for reuse and recycling have become impossible. The latter is the "material fate" of large amounts of construction rubble

Construction waste group	Materials	Occurrence in million tons	Recycling rate in %	Other recovery* in %	Disposal/ landfilling in %
Construction rubble	Concrete, bricks, tiles, ceramics	54.6	77.8	16.0	6.2
Road breakup	Bituminous mixtures	13.6	93.7	4.1	2.2
Construction waste based on gypsum	Construction waste based on gypsum	0.7	5.4	35.6	59.0
Soil and stones	Excavated soil, dredged material, railway ballast	118.5	10.2	75.5	14.3
Construction site waste	Wood, glass, plastics, metals and insulation materials, mixed construction and demolition waste	14.6	1.4	96.9	1.7

- Process technological advances in manufacturing that lead to higher raw material efficiency,
- More intensive use of products (Sharing Economy, but also smaller living spaces per capita, shared living areas, multi-family instead of single-family houses),
- Increased reuse rates,
- Extended product life cycles.

(*IRP*, 2020, *p. 37*).

If we assume lifespans of residential buildings of 50–70 years, it becomes clear that the buildings of the 1950–1970s are currently being demolished—and thus building materials, especially brick- and concrete rubble, are being processed, reused and disposed of in historically unprecedented quantities (UBA, 2013, p. 19). The material from a time of probably historically mass-intensive construction period is currently and in the coming decades thus being reintroduced into the "material cycle" (Fig. 6.4). The crucial question, however, is in which—i.e., how high-quality the recycling will succeed.

The sketched example of concrete is illustrative. The production of building materials is energy intensive. The importance of as high-quality recycling or reuse is increasing for all building materials, for reasons of expected resource shortages, increasing energy scarcity, stricter CO_2 emission costs. Reductions are driven by a mix of economic, ecological and environmental policy reasons (also the costs and regulations for landfill are increasing, climate balances count more in business calculation).

A detailed, excellent overview of the many relevant aspects of building material recycling is provided by the standard book "Recycling of Building Materials: Generation—Processing—Utilization" "by Annette Müller and Isabel Martins (Müller and Martins, 2022 published by Springer Vieweg). The statements of the following pages and Table 6.3 are largely based on this.

Table 6.3 Selection and addition JG, according to Müller (2018), p. 366f

Group of building materials	Reuse	Further utilisation/use
Ceramic building materials (Roof tiles, bricks)	Possible with damage-free disassembly	Cleaning e.g. of mortars, processing, crushing, for example into gravel
Natural stone	Mining and sorting, reuse of wall elements, cobblestone, natural stone slabs possible	Can be reworked, crushed, in extreme cases also into gravel or fine aggregates
Concrete	Hardly used monolithically; precast concrete parts, however, are difficult to transport	Concrete is shattered, crushed, ground into gravel, or as RC aggregate; fragments for anchoring purposes, such as dike construction
Steel	Possible if deformations and yield points are recorded; but reinforcing steel from reinforced concrete is difficult to separate cleanly	Cuttable, formable; limited reusability. RC capability very high through melting
Mineral binders	Lime, gypsum or cement mixtures are not directly usable because they have set	Grinding into flour and further uses possible
Non-ferrous metals	Aluminum parts, sheets of zinc, copper, can be removed reusable	Can be newly cut, shaped, processed
Solid wood	Ceiling beams, roof beams, wood paneling to formwork can be reused (this also applies to many wood-based panels, such as chipboard)	Cutting, planing, crushing, defibring, machining; bacterial, thermally or chemically digestible, as a raw material for the paper or pulp industries; combustible, energetically recyclable, possibly as compost usable

6.2.5 Spotlight on Building Materials

Masonry Rubble

Pure brick masonry is rarely found in the demolition masses (following explanations follow Müller, 2018, pp. 231–268). At least it is mixed with mortar residues, usually with many other construction waste fractions. Here we speak of *brick-rich masonry rubble* (secondary components: mortar, plaster). For the reusability, the brick content is decisive: "The maximum brick content should be at 95 mass-%, the average content at 80 mass%. Pure brick rubble can occur during roof re-coverings or can be obtained by pre-sorting from brick-rich masonry rubble. Production-internal brick waste also belongs to this category." (Müller, 2018, p. 232) Masonry rubble "is often used without or only with a minimum of preparation for the filling of mines, quarries and open-cast mines" (ibid., p. 262). Higher value uses would be easy to achieve. Technically "freed" from contaminants, recycled masonry can be used in road construction (where there is low load), in dam and path construction. Pure fraction masonry rubble can be crushed and also used as a raw material for the production of new bricks (Müller et al., 2013). Contamination with gypsum and correspondingly high sulfate contents often pose an obstacle to circular material management, for example because they lead to chemical reactions with ettringite formation or uplifts.

Brick Rubble

Even in the brick factory, production residues occur ("firing breakage"), which can be ground and reused for brick production:

> They act as lean agents and reduce the drying and possibly also the firing shrinkage. This is already practiced for a part of the firing breakage that occurs during production. A return is also possible for pure bricks from demolition. From raw mixtures, in which up to 50 mass-% of the clay was replaced by brick rubble, bricks of satisfactory quality were produced [...]. Unlike construction products made from standardized raw materials such as concrete, the level of substitution in brick production depends both on the type of product produced and the characteristics of the used primary raw material. For example, the substitution of 20 mass-% of a loamy clay by brick rubble was possible without deterioration of product quality. Very plastic clays like e.g. Westerwald clays could absorb up to 60 mass-% of the finely ground filler. Nevertheless, high-quality products with low porosity were producible (Müller, 2018, p. 253).

The decisive quality parameters for recyclability are particle sizes of the breakage material, loss on ignition, the sulfur and calcium carbonate content, and the proportions of the contaminants mortar and concrete (ibid.). The separation, cleaning and return of as large proportions of the masonry rubble into brick production would be ideal from the aspect of resource conservation, and a prime example of upcycling. The urgency of the project is underscored by the limited regional clay reserves (Fig. 6.4). Finally, brick sand can also be used as a sand substitute in concrete—however with decreasing strength (Fig. 6.5).

Concrete Rubble

The proportion of concrete reuse is still very low, extreme downcycling is the rule. Recycled concrete is rarely used. This cannot be explained technically alone. Because: "There is no significant difference between pure concrete aggregates and recycled construction materials from concrete rubble" (Müller, 2018, p. 200). The quality indicators of concrete rubble are roughly: raw, bulk density, grain porosity and water absorption capacity. Concrete from recycled concrete aggregates is standard-compliant, but only with restrictions usable in high-rise construction. There are also restrictions in civil engineering with special requirements for moisture classes, for example (Table 6.4).

As early as the 1990s, recycled concrete was used for "pioneer buildings" in Germany. These included the headquarters of the Deutsche Bundesstiftung Umwelt (DBU/eng.: German Federal Foundation for the Environment) in Osnabrück around 1994/1995 or the "Waldspirale" apartment building in Darmstadt from 1998 to 2000 (Fig. 6.6). However, these role models have inspired few successors:

> Although these demonstration projects have repeatedly proven that building with requirement-appropriate recyclates is possible, there was initially little progress in practical application. An initiative launched in 2009 by the Ministry for the Environment, Climate and Energy Economy of Baden-Württemberg demonstrated that concrete with recycled aggregates, which is produced in accordance with the regulations, is an equivalent building material and can have ecological advantages (Müller, 2018, p. 220).

The reasons for the preliminary failure of a mass deployment of these materials in the market are also seen in the higher costs of RC concrete. Low demand is also seen as a cause, with prices potentially falling due to scale effects of increasing production. The price surcharges are not necessarily large: examples for Germany indicate about 3% higher costs for the RC concrete (Hinz, 2018). But values of 25 percent and more cost difference to the disadvantage of the RC concrete appear in some cases as a realistic basis for calculation (Bachmann et al., 2018, p. 26). The legislator increasingly demands the use of RC materials from builders (see later sections). The EU legislator wants to make specifications for minimum recyclate

Table 6.4 Averages of the samples taken (number: n). *Selection JG, according to* Müller (2018, *p. 201)*

	Pure concrete aggregate	RC material from concrete rubble
Bulk density (kg/m³)	2678 (n = 34)	2596 (n = 13)
Raw density OD (kg/m³), Aggregates > 4 mm	2316 (n = 45)	2277 (n = 28)
Particle porosity, calculated (volume percent)	14.2 (n = 34)	14.5 (n = 13)
Water absorption, Aggregates > 4 mm (Mass percent)	5.7 (n = 44)	4.6 (n = 13)

Fig. 6.6 The Waldspirale in Darmstadt was completed in 2000. Photos by Jan Grossarth

shares. But this hardly improves the climate balance. In the Swedish study presented in the later life cycle assessment sub-chapter, the saving potential through concrete recycling was only estimated at just over 10%. Because so-called RC concrete saves the breaking of new natural stone, but not significantly the particularly climate-damaging cement (Mostert, 2021). It is also not suitable for the production of pre-stressed concrete and lightweight concrete components (Müller, 2018, p. 212). But at least large amounts of material can be saved, as scientifically accompanied practical examples show (Kreft, 2016).

Limestone Rubble

Pure limestone aggregates from demolition are used in plant cultivation due to their high porosity as substrates, for the greening of flat roofs, in road construction as gravel bearing layers or in frost protection layers, but also as a component of recycled aggregates in concrete production. The addition of 10–30 mass percent is recommended for use. "Pure limestone recyclates, which are produced as cutting waste during the packaging of limestone construction sets, are already being recycled in small proportions in internal production cycles" (Müller, 2018, p. 256).

Gypsum

Gypsum is actually—like clay—a fully recyclable building material. But adhesions, mixtures and impurities prevent recycling in practice. The majority must be deposited. For this, it is often transported to Eastern Europe, where the disposal costs are lower. Gypsum fractions are—as already described—a major obstacle to the recycling of construction waste, as they reduce its structural suitability. They contain contaminants such as the mentioned sulfates as well as chlorides and other so-called leachable salts. These interact with other contaminants such as asbestos or PCB (UBA, 2013; Müller, 2012). A higher recycling of the gypsum itself is also urgent because with the German coal phase-out, a by-product for the production of FGD gypsum will be eliminated—synthetic gypsum from flue gas desulfurisation; currently, 55% of the gypsum demand is covered by this (Riechert, 2022). A selective separation of gypsum boards before demolition can significantly reduce the contamination of the construction waste.

Copper, Iron

The history of prognostication is also rich in errors with regard to the reserves of copper and their availability. In the "Limits of Growths" of the Club of Rome, it was said that 1.5 gigatons of copper could still be mined; since then, around 0.7 gigatons have been mined—but by 2020, the estimate was already more than five gigatons of copper still to be mined. In the upper earth's crust, a total of 60,000 gigatons of copper resources are suspected (Unnerstall, 2021, p. 129). Metal extraction is also energy-intensive. Therefore, it is likely to reach its limits. Even today, the metal content of mined copper ore is often lower than decades ago (see the preface of this book). A shortage of iron will occur earlier. Here, the amount already installed will increasingly—and clearly from the 2050s onwards—become a globally more important raw material source than pig iron (Hiebel and Nühlen, 2016).

Waste Wood

In theory, a lot can be done with waste wood—even beyond the outlined biotechnological processes of lignocellulose utilisation—without immediately burning it. It can be polished and used as interior for luxury hotels in alpine ski resorts. Sometimes fresh wood is even modified for such uses to look like old wood. Its use in the so-called first German recycling house, which was built in 2019 in Hannover Kronsberg predominantly from recycling materials in the interiors (for example for paneling), was diverse. However, it failed in the outdoor area due to the formation of mold on waste wood from sauna benches (Gundlach, 2020).

This area is comprehensively regulated by law. The Waste Wood Ordinance generally allows the following ways of recycling for waste wood:

- Material recycling:
  - Processing of waste wood into wood chips and wood shavings for the production of wood-based materials
  - Extraction of synthesis gas for further chemical use
  - Production of activated carbon/industrial charcoal

- Energy recovery:
  - Recovery of waste wood in the sense of § 3 paragraph 23 in connection with the process R 1 of Annex 2 of the Circular Economy Act.

Particularly problematic is the reuse of waste wood of waste wood categories A III and A IV (the latter is "waste wood treated with wood preservatives, such as railway sleepers, utility poles, hop poles, vine stakes, as well as other waste wood, which due to its pollutant load cannot be assigned to waste wood categories A I, A II or A III, excluding PCB waste wood"). Pollutant-contaminated woods of category A IV may only be thermally recycled. Especially prefabricated wooden houses from the 1970s can be referred to as special waste landfills in this respect. Their chipboards were literally soaked in formaldehydes. These still emit into the room air today. This can be harmful to health when windows are closed. In Austria, it is said, the disposal of a prefabricated house from the 1970s is more expensive than the disposal of a new brick house (Teischiner, 2023).

Waste wood plays a virtually very small, but potentially a relatively large role as a raw material for timber construction (Türk, 2014, p. 341). Waste wood makes up a large part of the German wood supply: after all, 11.5% (Acatech, 2020). The total amount of waste wood is expected to increase significantly by 2030 (Wastemonitor, 2019). Experts see great potential for an increase in material use. 78% of waste wood currently goes into energy recovery, only 15% into material use and 6% are disposed of (UBA, 2020, p. 62).

The proportions of material use of waste wood vary depending on the specific wood-based building material. Among the building materials, it is in Germany mainly the chipboards that have the largest share of waste wood components. In Germany, this is around 20%. Italy achieves significantly higher values around 90%, the Scandinavian countries Finland and Norway zero: Denmark, on the other hand, around 60%; the material recycling rates of waste wood are thus usually significantly lower than in comparison to plastic (Teischiner, 2023).

Reuses of wooden components such as window frames or even structural beams have been scientifically proposed, and there have been economic potential analyses for such projects; in a final report of the project "CaReWood" for example, it was stated that the technical foundations have been laid (Risse, Richter 2018). But as far as marketability is concerned, the actual waste wood price and also the achievable sales price of such circular wood building materials proved to be essential variables that are likely to determine success (ibid., p. 120f.). Because of the uncertainties about the developments of these prices the industry remains cautious. But:

> Despite the general reservations on the part of the industry, there are open-minded and interested wood-processing companies that have already taken up the idea of full wood processing for the production of their products. A next step would therefore be the collaboration with companies for a more detailed and practical elaboration of the CaReWood process, e.g. in the form of a pilot plant. For the economic success of cascade use, in addition to the industrial, societal acceptance of circular products is necessary (ibid.)

Table 6.5 Suggestions for value chain innovation in the timber industry (after Coenen et al., 2022)

a. Linear supply chain management	b. Closed-loop supply chain management
• Maximisation of economic value creation • Economic and technical product life cycles • Finite usage cycle of products • Waste management	• Maximisation of "multilayered" value creation (sustainability values) • Technical product life cycles • Cascading, interconnected usage cycles of several goods

Thus, the implementation of the technical possibilities here again at the limits of acceptance and willingness to pay. How laborious it is, is also shown by the fact that such "visions" for a cascading and predominantly material use of waste wood were already formulated and scientifically accompanied in earlier decades—but without resounding success (Teischinger, 2009, p. 9).

In the context of climate policy ambitions and goals, an increasing number of companies are once again trying to implement such business models of closed wood material cycles. In a Dutch project, they talk about "Timber Loops" (Coenen et al., 2022). The multiple barriers that previous attempts encountered, led to scientists looking for principles for a *Closed-Loop Supply Chain Management* (CLSC) of the *supply chains.* In doing so, they called for nothing less than a completely new way of thinking and acting in the wood processing industry—a "mental shift", "a new way of thinking and acting"—or, as it is also stated in the study:

> Achieving CLSC management requires a structural change in the way supply chain actors think and act. Although a growing number of supply chain actors participate in the discourse on the circular economy, so far successful breakthroughs towards CLSC management are rare.

This frustrating finding was supplemented by a contrasting comparison of the characteristics of a classic-linear and the *circular timber construction management* (Table 6.5).

But do many companies in the established wood value chains have an interest in such profound change? Sawmills, for example, make very good money from the production of heating pellets—why should they support the demand for higher proportions of material (old) wood use or help implement it? Companies also often adorn themselves with sustainability slogans that do not live up to the serious claim of the concepts. Legally unprotected terms such as "varietal purity" were already attached to wood materials in the 1990s by manufacturers as a distinguishing feature, for example against Italian competitors (Teischinger, 2023).

Wastepaper

The use of wastepaper significantly improves the water, energy, and emission balances of paper production and conserves wood reserves (Türk, 2014, p. 190ff.) The wastepaper recycling rate in Germany in 2020 was more than 90% (Teischinger,

2023). The 100% is considered technically unattainable, as certain plastic components are contained in the wastepaper. Wastepaper can be of increasing importance for the construction industry, especially as a source of lignin and cellulose.

6.2.6 Separation Techniques

The many barriers to people and institutions driving the project of transformation sluggishly or not participating at all are the inertial forces of the "linear model" of the economy. The importance of the technological developments of separation and material techniques is large and often not given enough attention in the construction sustainability discourse. Because they contribute to making higher-quality new things from waste materials—and strengthen a circular economy, without the need for far-reaching behavioral changes in people. A more precise and multi-stage technical separation of the construction waste fractions appears to be a key to a circular management of the "resources" of the *Urban Mine*. The development of energy-saving.

- chemical,
- physical,
- mechanical,
- chemico-physical,
- optical, technological separation processes.

is therefore significant (Mettke et al., 2019, p 123). Here, a problem is the repeatedly mentioned large proportion of gypsum breakage—but it is also an example of separation technical solutions. Regarding the separation of harmful gypsum fractions, experts saw a practical, technical solution several years ago—at least for the separation of gypsum from lightweight concrete. A mechanical process developed at the University of Weimar, among others—the night separation with a so-called *attrition drum*—led to a 90–95% recovery of the lightweight concrete being possible (Baufachzeitung, 2013).

The Institute for Applied Building Research in Weimar (IAB) tested the color of the building material fractions (red bricks, gray concrete, brown or dark gray wood) as a criterion for sorting coarse aggregates by *sensor-based methods* in the form of infrared cameras, which have been successfully used in recycling in the paper or glass processing industry for years (Müller, 2018, p. 125ff.). The goal is to separate the construction waste from the contaminants asbestos, plastic, paper—as well as gypsum and to recover pure fractions of brick, (lightweight, porous) concrete, sand-lime brick and the gypsum itself. The process is still considered too expensive and currently there are few investment incentives for waste recovery companies. This could change with increasing demand for RC material, for which an end to the "waste property" would be necessary from an environmental law perspective. "The optical sorting results in an almost 100% pure gypsum fraction" (Müller, 2012).

6.2.7 Mortarless Solid Construction with Sand-Lime Brick

A replacement of mortar by steel tension connections in sand-lime brickwork is a researched and suitable alternative that meets the demand for separability and reusability of the masonry bricks. It was investigated under the acronym "Re-MoMab" at the TU Dresden (Youssef et al., 2019). Its findings are: Massive construction with pure-grade disassembly and reuse is possible, and this applies to all essential components of the wall structure and also the dry ceiling. The sand-lime brick could be reused in its pure form after deconstruction. The construction time is shorter, construction is possible even at low winter temperatures, and the bricks can be installed mechanically (Berninger, 2021). Research questions that are still open mainly concern the economic viability, realistic paths for the development of return systems or the development of construction financing systems that take into account reusability (ibid).

6.3 Case Study: Sustainable Concrete Pathways

The climate-damaging effect of cement—the particularly critical component of concrete in this respect—is *more than twice as high* as that of global air traffic (Inside Climate News, 2022; Fig. 6.7). About *8 percent* of global CO_2 emissions are due solely to cement production. If the *cement industry were a country*, it would be the third largest *climate sinner* in the world after China and the United States. This is because lime has to be heated to cement clinker at around 1450 °C, which costs a lot of (fossil) energy, and the chemical process of calcination releases a lot of CO_2. But there are some approaches to making concrete more climate-friendly in production or use; the alternatives are often still in the research phase.

6.3.1 Biogenic Additives

Renewable resources do not have to be played off against concrete. They can also be mixed in. This usually worsens properties such as compressive and tensile strength, but the climate balance improves. Examples are *paper ash, hemp straw* or Miscanthus fibers; applications in 3-D concrete printing are also possible (Acikel, 2011; Luhar et al., 2020). For a concrete building material with straw and wood ash, there was even once a business start-up in Austria, perhaps too early. Because the so-called "*Mixolith*" (Berger et al., 2008) apparently disappeared from the market again. The flour of the *prickly pear* as a binder even increased the compressive strength of the concrete. However, this will not save the world's climate. Prickly pear flour could only replace 0.5–1% of the *cement powder* (Aquilina et al., 2018); other studies indicate values up to 4%. The prickly pear can be seen as a prime example for a bioeconomic multiple use of a renewable raw material, since dyes can also be extracted from its fruits and the fruits are also edible. They can be meaningfully integrated into agroforestry systems. Another question, however, is

Fig. 6.7 Cement builds high: Resource-intensive new construction of the skyscrapers "Four Frankfurt". Photo by Jan Grossarth

whether a biogenic admixture in cement is desired at all. Because it leads to a difficult separability of the technospheric (mineral) and biogenic (renewable) building materials.

6.3.2 Simple, Separable and Long-Lasting Construction

So wouldn't it be better to design pressure-resistant concrete parts so that they last "forever"? This would correspond to the ideal of "*Cradle to Cradle*"—components that are reused forever. The longer the lifespan of a building, the better the life cycle assessments. And if concrete ceilings or beams—and this even in the case of *steel components*—are screwed, they can be separated without damage and reused. In this respect, concrete has a greater future the simpler the construction of tomorrow will be. A concrete house in Bad Aibling of the mentioned *research project* "Simple Building" of the TU Munich showed decent emission values because the scientists relied on pure construction. That basically longevity can be the great strength of concrete, visitors to Rome know: The Pantheon has been standing for almost 2000 years and was built from *Opus Caementitium*, a distant relative precursor of today's concrete (Sect. 6.4).

6.3.3 Textile Reinforced Concrete or Carbon Concrete

In the post-war decades, reinforced concrete became the building material of choice for both ceiling and bridge construction. But it can suffer damage when water penetrates. Therefore, many *bridges* now have to be demolished and rebuilt. The longevity can greatly increase if the risk of corrosion is reduced by new materials. Marcus Ricker from the Rhein-Main University of Applied Sciences has examined with co-authors how a concrete performs that is reinforced with *textile* fibers (Ricker et al., 2022). He used, for example, hemp fibers soaked in epoxy resin. The first result: The compressive strength is higher compared to unreinforced concrete. The material may not be suitable for bridge construction, but it could be suitable for ceiling construction. In addition, the climate balance improves. A working group of a The Fraunhofer Institute (2018) also works with textile reinforcements made of *flax*. In all these cases, however, the use of synthetic resin for hardening the reinforcing fibers represents an ecological disadvantage in terms of separability and reusability at the end of the life cycle. This objection to the complex separability also applies to carbon concrete (Scope et al., 2022). An advantage of textile concrete is that thinner concrete covers are possible because there is no risk of corrosion, which allows for slimmer construction methods and material savings (Adam et al., 2020). Disadvantages are that individual approvals are still needed for use, and the production of the reinforcements of carbon concrete is very CO_2-intensive.

6.3.4 Self-Healing Concrete

A research project at the Technical University of Munich takes a different approach. The idea: A concrete that repairs its cracks *itself* and thus lasts even longer (Baunetzwissen, n.d.). They use bacteria that produce lime deposits. They thrive as soon as the concrete forms cracks and moisture penetrates—a relevant application of bioeconomy of buildings. The Munich University of Applied Sciences is researching something similar. Here they are looking for spores that can ideally survive decades or centuries without water and *nutrients* (HS Munich, 2021). This application promises to increase the longevity of concrete facades. In addition to bacteria, there are other approaches to giving the concrete a protective shield—for example made of plastic (polymers) or minerals. For example, plastics of the classes polyurethanes and polyacrylates have been used for the healing of *concrete cracks* (de Belie et al., 2018). "Self-healing can extend the concrete lifespan by a significant amount, but leads to much higher production costs," summarise *scientists* at the University of Rio (da Costa et al., 2021).

6.3.5 Lightweight Concrete, Gradient Concrete and Wood-Hybrid Ceilings

For decades, a lot of concrete was used in many places from today's perspective unnecessarily—based on the then state of knowledge and the prevailing priorities, however, understandable—a lot of concrete was installed. As long as the energy was cheap, such "material wastages" could prevent liability problems. Just by thinner walls, as well as airier concrete materials, the climate balance could be improved. This idea is primarily represented by Werner *Sobek* from the University of Stuttgart. One of his *calculation examples*: The so-called gray energy—all energy that is in the building substance—of a 20-story high-rise could be more than halved, if instead of conventional reinforced concrete less material in the facades and so-called *hollow core slabs* were used. The majority of the savings would be on the concrete; a smaller part would be on *CO_2-reduced reinforcing steel*. If you now also use so-called *Wood-hybrid ceilings*, which consist of wood and concrete, improved the balance from 433 kg to just 94 kg of CO_2 equivalents per square metre of living space. Another idea of Sobek's: The use of *gradient concrete*, which contains a lot of air and is modeled after the structure of bones. It offers great potential for savings. After all, the climate impact of cement is not just a matter of material. For example, politics would have to stop designating rural residential areas for single-family homes. Because their climate balances are—including the necessary development—often *worse* than those of urban living in multi-story houses (Weidner et al. 2021; Wörner et al. 2016). Moreover, the attractiveness of living in the countryside is reduced by this sprawl.

Carbon Capture and Storage

The big unknown variable in this calculation is the development of Carbon Capture and Storage (CCS). Will it be possible to capture a large part of the CO_2 emissions directly at the cement plant and store them in deep layers of the earth? In *Norway* the first factory started doing this in 2021. Thus, half of its emissions were saved (MAN, 2021). Heidelberg Cement sees in this technology the way to become climate-neutral in the future. But still this is expensive. The cement price would have to *double* (Strunge, 2021).

6.3.6 Case Study: Lessons from Roman Concrete (Opus Caementitium)

Rome is the place where the building material concrete shows its two sides: durability and decay (Fig. 6.8). The durability can be seen for example in the Trajan's Forum in the ancient centre. There, Roman concrete artistry endures in many forms: As load-bearing walls, columns, inside and outside—exposed to the weather for almost two thousand years, but unshakeable. At their core, these walls consist of rock-solid brick rubble mix. They are a case of exemplary "upcycling" of construction waste, one would say today in sustainability jargon. The walls are clad on

Fig. 6.8 Two cement worlds in Rome: Long-lasting Opus Cementitium in the Trajan Forum, short-lived reinforced concrete near the beach. Photos Jan Grossarth

the outside with bricks or marble slabs. Inside they are glued together by the ancestor of concrete, the “Opus caementitium”. And at ancient places like the Trajan’s Forum, not only the durability of the ancient building materials is revealed, but also the variety of the building crafts of that time. In ancient Rome, there were—made by thousands of slaves—handmade bricks of many brands, recognizable by their own stamps, a quality competition.

Rome—or actually: all of Italy–is also a place where one can marvel at the sins of the cemented engineering art of the post-war decades. A short trip to Ostia, the Roman port, is enough for example. There you can see, in the first row by the sea, for example a concrete cast gas station from the 1950s. It is young, but already a ruin: Corrosion damage, the building inspector would say. Rusty steel structures wind out of the crumbling concrete mass, barriers prevent access. Nobody wants to know anything about this building anymore. Hundreds more of these stand along Italy’s coast, and inland, where the cement-strong bridges are suicidally breaking.

In the past, not everything was always better in construction. But the Concrete in the Latin world empire was already advanced in many ways. But why exactly? The corrosion damage and the annoying short lifespan of the engineering structures of the twentieth century are partly a specific issue of reinforced concrete, for example in connection with penetrating salt air and water. But beyond that, there are differences in the concrete recipes that matter. There must be something to the Roman cement, whose structures survive 2000 years? To this day, for example, the dome of the Roman Pantheon, built from Opus caementitium in 128 AD, still stands. Water still flows over ancient aqueducts, and even through underground pipes made of this material. The search for clues in ancient building material theory seems all the more important as sustainability and resource conservation in the “climate sinner industry” construction should also be understood as greater longevity. The exact recipe of the “Opus caementitium” still puzzles chemical science to this day. Even ox blood was added to this building material. A few years ago, researchers from the University of Utah found out how the added volcanic ashes strengthened the self-healing powers of the concrete—for example, when seawater penetrated into port facilities (Jackson et al., 2017). In a study in the journal American Mineralogist in 2018, they described which chemical reaction of seawater, lime from limestone, and volcanic ash was the reason for the durability. The ashes, so-called pozzolans, were finally, as laboratory tests proved, transported far from the place Pozzuoli near Naples and transported in ships. But that did not explain everything. Now science reports the next breakthrough in knowledge. Admir Masic, a professor of civil and environmental engineering at the American MIT, who was still researching at the German Max Planck Institute for Colloid and Interface Research in Potsdam a few years ago, has taken a closer look at the ancient "concrete" with his doctoral student Linda Seymour and others. Because to him, Masic, a detail of the previous building material historiography seemed too strange: Was it really only the pozzolanic—i.e., responsible for the strength of the concrete post-hardening—materials like volcanic ash that had caused the long durability? But Masic was interested in another detail. Under the microscope, he saw tiny minerals less than a millimetre long. These are called "lime clasts". So he and his researchers followed the trail of these lime clasts. These are microscopic particles of limestone, which originated from fragments of the skeletons of ancient corals or amoeba-like single-celled organisms. The scientists around Masic from the American Harvard University, with support from Italy and Switzerland, followed the trail of the lime clasts more closely, and published the results in the journal Science Advances (Seymour et al., 2023). The lime clasts, says the new study, gave the concrete self-healing ability. Such lime clasts are "not found in modern concrete formulations," Masic told the science service Eurakalert: "So why are they present in these old materials?" Until now, it was said that they had only gotten in there through mix-ups by sloppy workers. "If the Romans put so much effort into producing an excellent building material, according to precise recipes that were optimised over centuries—why would they then put so little effort into producing a well-mixed end product?" Microscopes, imaging procedures, and chemical mapping techniques helped in the search for truth. In the process, the researchers found out that the Opus caementitium was mixed wet, and hot. The secret of durability, it is said, lies in the hot mixing of the burnt lime powder with water. This sets a different reaction in motion. To prove their hypothesis, the researchers made samples hot mixed concrete in the Roman style with quicklime. They produced small cracks in the hardened material and found that these actually "self-healed". If no quicklime was present, this did not happen. The lime clasts thus formed a more brittle structure. This makes, after hardening of the "cement", this material reactive with penetrating water. Through the reaction, a calcium-saturated solution is created, which "heals" the cracks. A reaction with the pozzolanic ashes is added—thus the "old" explanation of the mystery of durability is not invalidated but supplemented.

Soon, Masic revealed, the recipe should be commercially used. Then it could also be of interest to companies like Heidelberg Cement. Masic talks about applications in 3-D printing of concrete structures. Thus, there is now another piece of the hope that even a climate-just living humanity in the future on the concrete will not have to do without. Because cement production causes about 8% of global greenhouse gas emissions, more than twice as much as global air traffic, and for that reason nothing is as sustainable as long-lasting construction.

6.4 Circular Building Design

The principles of disassembly-friendly construction (*design for disassembly*) differ from "conventional" planning, as cost and material wear calculations must extend over longer time horizons (Elbers, 2022). Product design as a discipline has historically emerged with the linear economy (*"take-make-dispose"*), but in this tradition products are explicitly understood as *time-bound* and of *limited lifespan*. The term "planned obsolescence" even refers to a deliberately implemented strategy of planned limited durability of consumer and utility items (Moreno et al., 2016).

From this, the planning claim of the *circular design* differs. Also the historical thread of the Dessau or Weimar Bauhaus is reinterpreted in the circular-economic sense. The historical model embedded the new possibilities of its time, to build industrially, primarily in the context of the *Social Question* as well as culturally, so the claim of an "ecological embedding" within the framework of the *New European Bauhaus* of the 2020s now includes the intention of a departure from linear design. In its place comes a comprehensive life cycle orientation in the sense of a consideration of repair and reuse possibilities. The planning of processes and product design would have to comprehensively orient themselves to the ideal of circularity. This affects many levels, such as:

- Material compositions
- Material connections
- Material durability
- Communication: Certification, Marketing, Dismantling instructions
- Business models: Rental instead of sale
- Urban planning (development plans, building area evidence).

Principles of Circular Design

Product, process design, and construction planning with regard to recyclability is often a prerequisite for their success (Fig. 6.9). The American Ellen McArthur Foundation (2022) has formulated three principles for *Circular Design*. They are, in short:

Fig. 6.9 Circular Design in Basel, Switzerland: "Weinlager" building with a circular corrugated metal facade by Esch Sintzel Architekten. Photo Jan Grossarth

- *Principle 1:* Preservation and enhancement of natural resources, controlled depletion of finite stocks, supplementation of material flows with renewable resources judiciously.
- *Principle 2:* Optimisation of resource yields. Products, components, and materials are used at all times—both in technical and biological cycles—with the highest benefit and circulate as long as possible.
- *Principle 3:* Reduction of harm to humans (in the areas of food, mobility, housing, education, health, and entertainment); minimisation of negative external effects through land use, air, water, and noise pollution, release of toxic substances, and climate change (EMA, 2022).

6.4.1 Component Exchanges

Digital trading platforms like *Concular*[1] expand the established component exchanges. Here, building parts such as bricks, window frames, glasses, or inventory are photographed and listed before demolition. Another trading platform is the Bauteilnetzwerk Deutschland (eng.: Component Network Germany),[2] and also the classified ad platforms like *Craigslist* are full of them. Component exchanges have existed for decades. But their history begins with the trade of decoratively valuable or high-quality antique components. Since 1992, the Unternehmerverband Historische Baustoffe (eng.: *Entrepreneurs Association for Historical Building Materials)* has existed as a merger of the dealers. It also acts as a broker to whom interested parties can turn—be it for the purchase or sale of old wooden floors, doors, bricks, etc.[3] Established historical building material dealers with large stock inventories traditionally exist in Belgium and the Netherlands. However, the use of historical building materials in new construction or renovation has limits in the energy saving regulations. Thus, older single or double glazed windows, but also old bricks, have significantly worse U-values than required by law (UBA, 2015, p. 54). However, innovative uses in the interior or in combination with energetic subsequent upgrades are possible.

6.4.2 Circularity Indicator

To measure the circularity of building materials, the *Material Circularity Indicator* (MCI) was developed by the *Ellen MacArthur Foundation.* It rates the product with a score between 0 and 1, with information such as the recyclate content, reuse options, and the lifespan flowing into the value. The calculation methodology is relatively demanding, suitable for companies and is explained step by step on the foundation's pages (EAF, 2023). The indicator was revised in 2023 to address issues of the shares of renewable raw materials and their mixing with to consider mineral resources more appropriately than before.

6.5 Policy for the Circular Economy in Construction

Several research institutes have also identified and evaluated measures for the WWF that can increase the circularity of building construction (Prakash, 2022, p. 64ff.). Here, the following were identified and evaluated in comprehensive assessment matrices with regard to values such as technological maturity, conflicting objectives, and entrepreneurial feasibility (mostly positively):

- Reduction of living and office space per capita,
- Dismantling planning, use of standardized building elements,
- Extended building lifespan,
- Local raw material procurement (stones),
- Take-back systems,
- Increased RC shares.

In summary, it is said that the circularity measures for building construction have "almost all a great potential for environmental relief". Cost savings and price reductions are expected primarily from a reduction in living and office space. However, these also raise—in connection with the living space prices for rent or purchase—the biggest societal acceptance questions. The establishment of take-back systems for insulation materials are also considered a relatively conflict-free approach.

From the companies' point of view, however, "moderate" to "high" cost increases are to be expected for "all other" measures. However, it is also stated here that the data situation regarding the questions examined is still "overall mediocre to poor". Available data often only based on case studies: "Extrapolating the data to the German construction industry is sometimes difficult."

For civil engineering, the

- Reduction of construction steel masses, such as concrete through slimmer design and
- Alternative cement binders.

are mentioned. They also promise "significant environmental relief potentials". Material savings are here *Win–Win*, as they lead to cost reductions for companies and price reductions. The use of alternatively produced *Celitement* (KIT Karlsruhe) alone promises a reduction of the greenhouse gas potential in this area by up to 50%, but it is three times as expensive as common cements. A concrete implementation is not barrier-free. Elbers (2022) mentions such obstacles for example in relation to the resource-saving planning of structures. The reuse of load-bearing elements is made difficult primarily by the necessary suitability tests, which can only be profitable if there is a sufficient number of pieces. But digital platforms can also record material quantities and bundle demand in this regard. In this sense, *Madaster* cooperates with the international construction engineering firm Arup. Possibly, overconsumption of materials in the structural area can even be worthwhile, as they increase the reusability: "For the structure, a system without composite systems can lead to larger cross-sections and thus to more resource consumption, at the same time, future recyclability is guaranteed." Work is also being done on take-back systems for individual building materials. Their conditions are being scientifically researched. An example is the aerated concrete as a niche segment of the concrete market. In a pilot study with the disposal company *Otto Dörner GmbH* from Hamburg and the aerated concrete manufacturer Ytong (*Xella*) researched how and in what quantities aerated concrete remnants from demolished houses or even from construction waste landfills are suitable for reuse as aerated concrete. The conclusion was that the purity of the material for the intended recycling was "sufficient" due to the combination of established and innovative separation processes such as metal separation, wind sifting, sink-float separation, manual resorting, pre-crushing in the crusher and sieving for defined grain size bands. Up to 15% of the thus processed old aerated concrete can be used in production (Kreft, 2016) (Fig. 6.10).

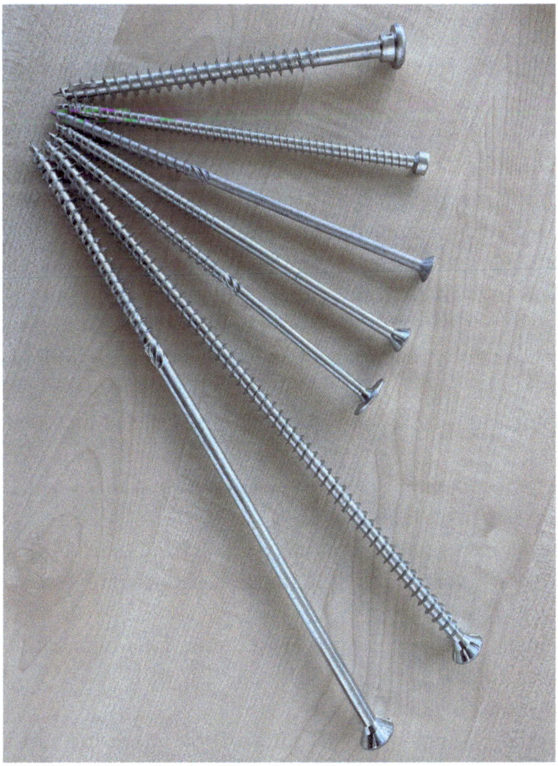

Fig. 6.10 Screws for many purposes: An unglued insulation board can be gently dismantled and reused. Photo by Jan Grossarth

6.5.1 Material Cadastres, Inventories and Passports

Digitisation can enable circular construction on several levels. Information about quantities, qualities, storage sites and connections of installed materials are necessary. At the federal level, a *material cadastre* is intended to obligate construction companies, to maintain this information for new buildings. Cities like Aachen (Aachener Zeitung, 2022) and Heidelberg (2022) have taken the lead with pilot projects. The German Umweltbundesamt (UBA/eng.: Federal Environment Agency) completed the pilot study "Mapping of the anthropogenic stock" in 2022 (Federal Environment Agency, 2022b) and is developing indicators that are also suitable for feeding regional material cadastres for regions and municipalities.

Material inventories refer to directories for individual buildings, *material cadastres* on the other hand for building stocks. The latter are therefore defined as

> Instruments that can describe the size and composition of material stocks in building inventories and their temporal changes. They refer to material quantities and material qualities

within defined spatial system boundaries or regions and can be expanded on the input and output side to include raw material and waste management aspects as well as material-induced effects (e.g. grey emissions) (Federal Environment Agency 2022b, p. 25f.)

Building material flows can therefore be found in the construction activity statistics or waste statistics. However, building type-related material key figures are necessary; in the mapping project of the UBA, the claim is that such data can then be "directly converted into raw materials (by considering building material recipes), waste categories (by assignment and further differentiation of waste categories) and grey emissions (by integrating life cycle assessment data sets)" (Ibid., p. 26).

According to which components such material cadastres should be systematized was worked out by expert commissions in 2023. One suggestion showed that it is not only about building materials in the structure. The list included:

- Foundations, substructures, exterior walls, interior walls, ceilings, roofs,
- Sewage, water, gas installations,
- Heat supply systems, heat generation systems, heat distribution networks,
- Ventilation systems, ventilation systems, air conditioning systems, refrigeration systems,
- Electrical systems, self-supply power systems, Communication, security and information technology systems (UBA, 2022, p. 69).

Building material passports provide information in the form of a mobile application about materials used in a building, their expandability and also the material value of a property. Madaster is the well-known example.[4] After recording the relevant materials used, the application also provides information about the monetary value of these building materials, their toxicity or their potential for reuse. It was used in the early 2020s in numerous "lighthouse projects" of circular construction.

6.5.2 Circular Urban Planning

Material savings cannot be achieved within a democratic economic order independently of changes in demand for resource-saving building products or buildings. If these are lacking, the project of the circular economy does not make significant progress. Therefore, *urban planning* plays a major role because it has the legal powers and the greatest "impact" to influence construction towards material conservation, regardless of the demand for types of housing (Weidner et al., 2022). This is done through criteria for development. Comparing the types of development single-family house, four- and six-story apartment building and 20- or 40-story residential high-rise, there are significant differences. The resource balance per square metre of gross floor area of the 6-story residential block in urban space is clearly superior to the other types of residential development—including the

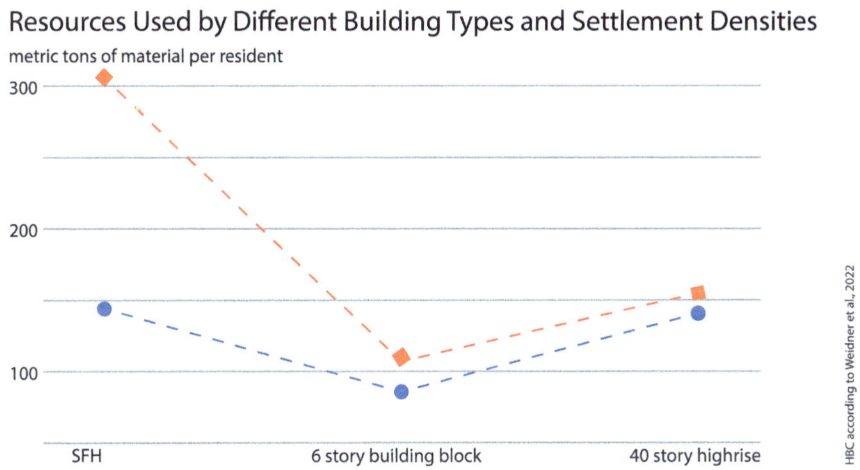

Fig. 6.11 The 6-story building block uses the least resources. (HBC according to Weidner et al., 2022)

necessary road infrastructure for development (see Fig. 6.11). Through a "radical optimisation of the reinforced concrete structure", the gray energy can also be significantly reduced for high-rise construction. Mineral hollow floor slabs reduce the consumption of concrete and also reinforcing steel, resulting in savings of 52 kg CO_2 equivalents per square metre gross floor area (ibid.). The use of recycled concrete is another way, as is CO_2-reduced cement, CO_2-optimised reinforcing steel and the partial use of wood. It should be noted as a limitation to the role of the public sector that harmonized building products must not be disadvantaged by authorities.

6.5.3 Pollutants: Case Study on EPS Panels

An interesting case are also return regulations for External Thermal Insulation Composite Systems (ETICS). Here there are due to contained pollutants still serious problems in terms of the reusability of dismantled panels; however, this is likely to change in the future (Heller and Flamme, 2020). Then return obligations or voluntary return systems would favor circular reuse. Currently, however, in Germany EPS panels are still almost exclusively incinerated. Even further away from the claim of a circular economy is Europe, where about a third of the EPS panels are deposited (Heller, 2020): EPS panels produced before 2016 contain, in addition to the "usual" adhesives and plaster residues according to chemical regulation *REACH* since then high doses of residues of the flame retardants *HBCDD* (German Environment Agency, 2017). This is now more strictly regulated. 12,000 tons of HBCD were used in Europe per year, mainly for insulation production (ibid. p. 3). Legal limits also regulate disposal. The panels of the years before 2016 must be exclusively incinerated, thereby becoming the flame retardant is destroyed and rendered harmless. Meanwhile, alternative fire retardants are in use (*Brominated Styrene-Butadiene Copolymer; pFR*).

6.5.4 A Utopia? Skeptical Questions About the Feasibility

The cycle, the circular is initially an image—the guiding principle of a sustainable economy—, a metaphor like the bioeconomy. But unlike the latter, the former has been significantly environmental law for decades. The image of the cycle is itself a symbol from the area of the organic, the medical: It reminds of the human circulatory system. *Circular Economy* is a vision, in its pure form certainly a utopian one. Because can it really succeed, a world without waste? Isn't the production of waste somehow also part of modern humanity, the waste to joie de vivre—irrationality, nonsense, ecstasy (Hanske and Sarreiter, 2023)? Isn't all this even a part of the modern *dream of urban life* (Moholy-Nagy, 1968), which is realised now and then? Lights in the night, concrete, glass, comfort? How can it be integrated into everyday action, constantly thinking that this is all in a planetary sense, "unhealthy"—the metabolism of land and soil on the one hand, and food and place of life on the other, in the industrialized civilisation disastrously separated? Or: How can the environmental legal framework be set so that a circular economy becomes possible at reasonable costs? These questions indicate the conflicts, that circular planning and ideals bring with them. Then it becomes a question of ethical balancing: between planetary *health*, social participation and private or entrepreneurial freedoms (also for irrational pleasures).

Nature knows no waste—or does it?

"Nature knows no waste." This sentence is often heard also in engineering Sunday speeches about sustainability. It can be found in the 2020s also in advertisements and announcements by retail chains like *Coop* or *Bauhaus*, in scientific writings from the field of environmental engineering and also material flow management since about the late 1980s. In the 2000s, this normative claim was popularized by authors like Franz Alt (2019). Nature was declared a "role model" in exactly this sense (Nachtigall, 1997). This shows how functional and concise this metaphor as orientation for resource conservation innovations was and is. However, the metaphor also implies that nature has a consciousness, a memory, is a person. It is interpreted in the sense that nature teaches us the smarter principles—so normative. The geographer Wolfram Mauser has critically examined the metaphor from an environmental historical perspective (Mauser, 2016). He therefore considers the idea of an economy without waste to be unrealistic and unhistorical. Ideas like the one of a closed water cycle go back to antiquity. On the other hand, "nature" teaches linearity: planets and stars have limited lifetimes, from the

Big Bang to the first life on earth, "material flows" have changed dramatically, black holes set "planetary boundaries" in their own way. Mauser looks at the action of cyanobacteria, which were the first to convert CO_2 and sunlight into biomass. Secondly, he looks at the foraminifera and first shells, which even created exoskeletons from it. The photosynthesis over the eons caused the carbon dioxide content of the earth's atmosphere to drop from about 100 percent to values well below one percent. This increased the oxygen concentration, which enabled more advanced life forms. The chalk relics, on the other hand, are the basis for today's limestone mountains. Mauser provocatively speaks of "Oxygen dumps" and sees in them "irreversibly changed" material flows—essentially purely "natural" phenomena, because living beings are oriented towards demarcation and reproduction, not towards waste avoidance. "The production of waste is an integral part of this optimising self-organisation process of life. The only recognizable norm by which this component is used is short-term success in reproduction. Categories such as waste avoidance, resource protection, circular economy and sustainability are not recognizable." (Mauser, ibid.) From a historical perspective, the fossil-based life forms of modern times are unparalleled in their energy hunger and redirect material flows in historically new ways, creating new "waste mountains" from CO_2 emissions to landfills. The question for Mauser is, how can these be used to serve life? A complete renunciation of "cheap energies" would consequently also have to result in urbanisation, social welfare and education systems being revised or "de-developed", as these have emerged with the "energy-hungry" economies and systems of political participation, argues Mauser. The special case of the CO_2 problem is that it has "dislocated" the waste problem. The idea of the circular economy is, thought reflectively, a sensible orientation value of resource conservation. At the same time, nature is not suitable as a norm giver from a historical perspective.

6.6 Examples of Circular Design

Beyond the theory (vision) and questions about its feasibility, and also beyond the complex legal, statistical and standardisation questions of the circular economy, there are a number of remarkable designs and buildings that have creatively and practically committed to the idea of upcycling. The range is to be indicated by three very different examples:

- In american Hank Louis (Utah), the residential Tiny Houses *Whitehorse* were built in 2000 as a student project 90% from used materials, including telephone poles as pile elevation, under which desert flying sand can blow through, or from railway sleepers, discarded stadium benches, old wood of various origins, from old ship pallets (Edwards, 2017).

- In Indian Maharashtra, the *Collage House* of architects Shilpa Gore-Shah and Pinkish Shah was built with facades made of scrap metal, stone tiles and pipe pieces and the window glass made of old glass, the materials were purchased from waste dealers from the Dharavi slum; chair covers are also made from textile waste, interior walls from pipe pieces from scrap yards. The architects are quoted with a critical statement on eco-balancing:

> With some concern, we observed that sustainability in construction was increasingly being evaluated on the basis of a point system. It was only about building environmentally friendly houses from appropriate materials. Unfortunately, the result was often shockingly banal and soulless.
>
> (Shilpa Gore-Shah)

- The office building *The Cradle* in Düsseldorf by architect Thomas Götzen was built in 2022 in a wood-hybrid construction method consistently according to Cradle-to-Cradle standards and accordingly certified. It stands on reusable V-concrete-supports, only C2C-certified materials were used for the construction. The supporting structure is processed with RC concrete, the wood is regional, all components are digitally stored in the Madaster building pass, by programming a digital twin in the BIM system in the planning phase, the gray energy was consistently minimized during construction (Handelsblatt, 2022).

6.7 Cradle-To-Cradle Certifications

The approach known under the brand name *Cradle to Cradle* (C2C) clearly differs from the political-planning approaches to resource conservation and minimisation of the ecological footprint. The central idea here is a strict separation of the cycles of the technosphere and biosphere (Braungart and McDonough, 2002). The private sector certificate of the *EPEA GmbH* based on this idea goes back to the work of its founder Michael Braungart (see preface). The *EPEA* was sold to the construction project company *Drees & Sommer* in 2019 (Drees & Sommer, 2022). In parallel, there is the *C2C-NGO*, which advises entire municipalities on the implementation of C2C models (C2C Construction, 2022).

C2C is primarily, also internationally, popular through the writings of Braungart. It has become a *lifestyle label*, and its sharp distinction from other building sustainability certificates is explained by its consistent focus on circularity, which is also characterized by the occasionally polemical, at any rate very pointed statements Braungart makes about the sustainability expert world and the eco-certified food industry. The former, he says in essence, optimise the wrong things, and the latter are quasi-religious belief systems that draw their energy from attributions of guilt, that humans are harmful to nature. However, according to Braungart, technology could even improve environmental quality. Instead of constantly striving to make the environmental harm of humans somewhat less negative, construction should become environmentally positive: Humanity should then "celebrate" its footprint (Braungart, 2016).

This is the high claim of the C2C certification. EPEA has already certified numerous construction products (C2C Cert, 2023; EPEA, 2023). Three core criteria for certification are:

- Reuse or reusability of materials,
- Dismantlability,
- Defined usage path beyond the product's lifecycle.

The assembly systems used are central to the evaluation. They decide on separability. The quick assembly system *Würth Varifix* is an example of a certified system (Kübler, 2021).

6.8 Historical Approaches to Gentle Demolition

Only the historically unprecedented availability of energy after the Second World War paved the way for the careless and wasteful handling of the building stock. Before that, the building materials were much too valuable. A historical example is the Parisian *Palais des Tuileries*. It was demolished in 1882. As usual, several demolition companies bid, and Achille Picard was awarded the contract. The city of Paris secured a large part of the materials, the natural stones—or entire arcades and wall sections—are still present in many buildings that were built during this time, such as museums or parks. Many private individuals also secured valuable building materials before the demolition. The country house *Château de la Punta* in Corsica was largely built from the demolition material of the Parisian Palais achieved (NZZ, 2010).

Even in the 1920s, a division of labor and craft-based construction material trade was established in the cities of Europe and America and closely linked to the demolition industry. In New York, it was not the builders who paid a demolition company money for "demolition" and "isposal"—as is common today. Instead, it was the other way around. The demolition contractor paid. Jacob Vol was one of these entrepreneurs in the metropolis on the American *West* Coast. He outbid other competitors for deconstruction licenses, from private property owners or the city. A deconstruction project in 1921 made headlines. It was the city's outstandingly large skyscraper at the time of construction—in 1910—the *Gillender Building*. Just twelve years later, it was considered too small for the valuable location, with even taller buildings having been erected all around. In just 45 days, the numerous workers dismantled it for Jacob Vol.

The short lifespan of this modern building shocked contemporary observers. The rapid deconstruction of the *Gillender Building* "thus also marked the end of the common practice of directly reusing building materials" (Cortés et al., 2021, p. 80). Now a large public also saw how quickly even modern buildings go out of fashion and how limited their lifespan has become (Fig. 6.12).

Historical photographs and film footage of deconstruction projects from this time—such as a time-lapse film of the *Star Theatre* in New York—show that a building was carefully dismantled from top to bottom like an anthill. Window frames and glasses were removed, brick by brick, as well as the solid wood of the roof structures. The deconstruction companies maintained large building material warehouses. The used building materials even had advantages. The wood was already dried and builders could be sure that it would not shrink anymore.

Almost everything was used, as a *Master Thesis* at the *Columbia University* shows (Arlotta, 2018). Bricks and natural stones were dismantled and cataloged in such a way that walls could be rebuilt in the same way. The bricks and natural stones were carefully cleaned. Since they had not used hard Portland cement, but lime mortar, this was possible. Soft-fired bricks, which were no longer suitable for this, were used for the construction of non-load-bearing interior walls. Wooden planks were reused, and fragmentary plank floors were usually not burned, but used for chipboard production. Nail by nail was carefully pulled to not demolish the wooden planks. Rubble was used for the foundations of new buildings. Entire window frames and window glasses were reused in new construction projects. Particularly valuable were marble or granite stones or slabs.

The demolition work was probably hard, and it would be questionable to what extent it could still be financed at all under the conditions of socially acceptable minimum wages with regard to the goal of moderate costs for used building materials in our time. But could machines help? How could construction planning take into account a possible simple later deconstruction? In the sense of these questions, looking into history is not only enlightening but necessary.

Oogstkaart in the Netherlands

As a used building material marketplace in the Netherlands, Oogstkaart is a successful example that facilitates Urban Mining. Several hundred used building materials or parts can be found there, marked on Google Maps. The platform was created in connection with the renovation of a residential building in the harbor district of Rotterdam in Wolphartstraat by the architecture firm Suoeruse Studios, which imposed the mandate to plan according to a strict hierarchy of the following criteria: Firstly, components and building materials had to be reused. If this was not possible, renewable materials were to be used. Thirdly, recycling material was to be used, which had to come from as close a radius as possible. The exact transport routes of the materials were documented. Only in fourth place was it allowed to use new, but once reusable building materials.

www.delcampe.net kd7850

Fig. 6.12 Turning point: The demolition of the Gillender Buildings in New York 1921 after only eleven years. Illustration Delcampe

The abrupt end of this industry was due to technical innovations, namely the wrecking ball and the bulldozer. This happened from the 1940s onwards (Arlotta, 2018). They changed the economics of demolition, demolition was very cheap to have, and at the same time, mass building materials became increasingly cheaper due to the decline in energy prices. It was also now easier to transport construction waste with electrified trucks, and later to cut and process it instead of dumping it. The "American dream" of liberation from hard bone work came true: "Americans were eager to leave the resource-strapped, penny-pinching war years behind, and the pendulum swung dramatically towards conspicuous consumption" (Ibid.)
The invention and spread of the *bulldozer* from the 1950s not only revolutionized the demolition industry, but also the building materials industry, trade, and pushed the used building materials markets into the niches of hobbyism. With the bulldozer, a *Culture of Clearance* was introduced, as the cultural historian Francesca Russello Ammon wrote (Ammon, 2016). It became "normal" to "clean" the area quickly and cheaply by mechanical demolition before starting construction and to build anew on the "cleaned area". A corresponding division of labour developed in the construction and demolition industry. In the 2020s, the question would be how, given the high division of labor and the established "value chains" in these areas, a focus on a gentle and contemporary demolition culture can succeed (see introductory actor sketch in Sect 2.3).

Notes

1. https://concular.de/.
2. www.bauteilnetz.de.
3. https://www.historische-baustoffe.de/.
4. https://madaster.de/.

References

Aachener Zeitung. (2022). *Neue Kita in Breitbenden.* https://www.aachener-zeitung.de/lokales/aachen/aachener-politik-fasst-beschluss-fuer-neue-kita-in-breitbenden_aid-83360913Kiefhaber.

Acatech. (2021, Ed.). *Urban mining.* Acatech Horizonte.

Acatech (Eds.). (2021). *Urban Mining.* Acatech Horizonte.

Achatz, A., et al. (2021). *Kreislaufbauwirtschaft.* Project Final Report, Federal Environment Agency.

Acikel, H. (2011). The use of miscanthus (Giganteus) as a plant fiber in concrete production. *Scientific Research and Essays, 6*(13), 2660–2667.

Adam, V., et al. (2020). Experimentelle Untersuchungen zur Verstärkung von Brückenfahrbahnplatten mit Textilbeton. *Beton-und Stahlbetonbau, 115*(12), 952–961.

Allgemeine Bauzeitung. (2023). *Gutex führt Rücknahmesystem ein.* https://allgemeinebauzeitung.de/abz/nachhaltige-kreislaufwirtschaft-gutex-fuehrt-ruecknahmesystem-fuer-paletten-ein-48736

Alt, F. (2019). *Die Alternative: Plädoyer für eine sonnige Zukunft.*

Ammon, F. R. (2016). Bulldozer. Yale.

Aquilina, A., et al. (2018). The application of Natural Organic Additives in Concrete: Opuntia ficus-indica. *IOP Conference Series: Materials Science and Engineering, 442*, 1.

Arlotta, A. I. (2018). *Locating heritage value in the reciprocal relationship between preservation and waste reduction.* Columbia University Masters of Science in Historic Preservation.

Bachmann, V. et al. (2018). *Recyclingbeton im Bauprozess.* Student work at ETH Zurich

Baufachzeitung. (2013). *Trennverfahren.* Accessed on 3.5.2022. https://baufachzeitung.com/recycling-verfahren-zur-trennung-von-leichtbeton-und-gipsputz-entwickelt/2013050720/

Baunetzwissen (n.d.). Selbstheilender beton. Accessed on 23.2.2023. https://www.baunetzwissen.de/beton/tipps/forschung/selbstheilender-beton-5402460

Berger, W. et al. (2008), *Entwicklung des ökoeffizienten Alternativbaustoffs "Mixolith" unter Verwendung von Reststoffen biogenen Ursprungs als Bindemittel: Ein Projektbericht im Rahmen der Programmlinie "Fabrik der Zukunft"; Impulse Program Sustainable Economy.* Federal Ministry for Transport, Innovation and Technology

Berninger, M. (2021). *Conversations and email exchange with JG; the graduate of Biberach University had written his unpublished bachelor's thesis on this construction method and plans to build such a house, Biberach.*

BGR. (2020). *Vorkommen und Produktion mineralischer Rohstoffe—ein Ländervergleich*, p. 18.

BMWI. (2019). *Rohstoffstrategie der Bundesregierung.*

BPIE. (2022). *Bausteine einer Lebenszyklusperspektive für mehr Klimaschutz und Ressourcenschonung im Gebäudesektor—Der Deutsche Rechtsrahmen und gute Beispiele aus der Praxis*, n.p.

BR. (2022). *Bremse im Bauhandwerk: Gips wird knapp—und teuer.* https://www.br.de/nachrichten/wirtschaft/bremse-im-bauhandwerk-gips-wird-knapp-und-teuer,T0H3VLG

Braungart, M. (2016). Learning to celebrate our human footprint. *RCC Perspectives, 3*, 27–32.

BVSE. (2020). *Statusbericht Kreislaufwirtschaft.* https://www.bvse.de/dateien2020/2-PDF/01-Nachrichten/01-bvse/2020/November/Statusbericht_der_deutschen_Kreislaufwirtschaft_2020.pdf

C2C Buildings. (n.d.). n.t. c2cbuildings.net.

C2C Cert. (2023). *Product database.* https://c2ccertified.org/certified-products-and-materials

C2C Bau. (2022). *Imprint.* https://c2c-bau.org/impressum/

Coenen, J., et al. (2022). *Expediting the implementation of closed-loop supply chain management: A facilitated case study on re-using timber in construction projects.* Circular Economy and Sustainability, 3(1), 1–32.

da Costa, V. C., et al. (2021). Core-shell system based on gelatin and poly (Vinyl Alcohol) (PVA) for concrete self-healing applications: synthesis, characterization, and optimization. *Macromolecular Symposia, 398*, 1.

Daxbeck, H. et al. (2009). *Urban mining—potenziale und handlungsoptionen bei der erschliessung von materialien aus dem bestand; ressourcen management agentur; vortrag auf der fachtagung ressource 2009, Ressourcen- und Recyclingstrategien—von der Idee zum Handeln*

de Belie, N., et al. (2018). A review of self-healing concrete for damage management of structures. *Advanced Materials Interfaces, 5*, 17.

DIN. (2022). *Circular economy.* Accessed on 20.1.2023. https://www.din.de/de/forschung-und-innovation/themen/circular-economy/anwenden/circular-economy

DIN. (2023). *Deutsche normierungsroadmap circular economy.* Accessed on 20.1.2023. https://www.vdi.de/fileadmin/pages/mein_vdi/redakteure/publikationen/Normungsroadmap_Circular_Economy.pdf

Drees & Sommer. (2022). *C2C.* Accessed on 31.4.2023. https://www.dreso.com/de/dreso/nachhaltig/cradle-to-cradle

EAF. (2023). *MCI.* Accessed on 30.1.2023. https://ellenmacarthurfoundation.org/material-circularity-indicator

Edwards, A. (2017). *Upcycled homes: Unique, innovative & sustainable living.*

Elbers, U. (2022). Ressourcenschonendes Bauen—Wege und Strategien der Tragwerksplanung. *Bautechnik, 99*(1), 57–64.

EMA. (2022). *CE introduction.* Accessed on 29.12.2022. https://ellenmacarthurfoundation.org/topics/circular-economy-introduction/overview

Entsorgungsmagazin. (2022). *RC-Siegel*. Accessed on 23.1.2023. https://e-mag.press/siegel-fuer-recycelte-baustoffe/

EPEA. (2023). *Leistungen*. Accessed on 20.2.2023. https://epea.com/leistungen/industrieprodukte/bauen-und-moebel

EU. (2011). *BauprodukteVO*. Accessed on 9.2.2023. https://eur-lex.europa.eu/legal-content/DE/TXT/?uri=celex%3A32011R0305

EU. (2020). *Renovierungswelle*. Accessed on 2.4.2023. https://eur-lex.europa.eu/legal-content/DE/TXT/HTML/?uri=CELEX:52020DC0662&from=DE

EU Commission. (2020). *Ein neuer Aktionsplan für die Kreislaufwirtschaft*. Für ein saubereres und wettbewerbsfähigeres Europa. Accessed on 9.2.2023. https://eur-lex.europa.eu/resource.html?uri=cellar:9903b325-6388-11ea-b735-01aa75ed71a1.0016.02/DOC_1&format=PDF

Eurostat. (2021). *Waste statistics*. https://ec.europa.eu/eurostat/statistics-explained/index.php?title=Waste_statistics

Federal Environment Agency (2022b). *Die Nutzung natürlicher Ressourcen. Bericht für Deutschland 2022, Ressourcenbericht für Deutschland 2022.*

Flamme, S., & Krämer, P. (2010). *Von der Abfallentsorgung zum "Urban Mining"* Vortragsfolien des Vortrags Urban Mining—Fachkongress für urbanen Umweltschutz, Meinungen, Konzepte, Zukunft, Iserlohn.

Flamme, S (2023). Schließung von ressourceneffizienten Produktkreisläufen durch neue Geschäftsmodelle—RessProKA. In *Lecture given at the 1st international Wood BioEconomy Forum (WBE)*. Rosenheim University of Applied Sciences.

Frankfurter Allgemeine Zeitung. (1993). *Töpfer warnt vor eine Atempause in der Umwelt- und Abfallpolitik*, p. 15.

Fraunhofer Institute. (2018). *Bauen mit Textilbeton aus nachwachsenden Rohstoffen*. Accessed on 23.11.2022. https://www.fraunhofer.de/de/presse/presseinformationen/2018/oktober/bauen-mit-textilbeton-aus-nachwachsenden-rohstoffen.html

German Environment Agency (2017). *Hintergrund//Dezember 2017*. Retrieved: 28.12.2022.

Global Footprint Network. (2019). *Ecological footprint*. Accessed on 12.1.2023. https://data.footprintnetwork.org/

Gundlach. (2020). *Deutschlands erstes recyclinghaus am kronsberg*. Accessed on 16.4.2023. https://www.gundlach-bau.de/zuhause-finden/infos-fuer-interessierte/mietinteressiert/referenzen/recyclinghaus

Handelsblatt. (2022). *Bauen ohne Abfall: The Cradle*. Accessed on 23.6.2022. https://www.handelsblatt.com/unternehmen/mittelstand/familienunternehmer/thomas-goetzen-bauen-ohne-abfall-so-errichtet-interboden-ein-recycling-buerogebaeude/28239178.html

Hanske, P.-P., & Sarreiter, B. (2023). Berlin.

Haque, M. S. (2019). Sustainable use of plastic brick from waste PET plastic bottle as building block in Rohingya refugee camp: A review. *Environmental Science and Pollution Research, 26*(36), 36163–36183.

Harris, S. et al. (2023). Climate mitigation from circular and sharing economy in the buildings sector. *Resources, Conservation and Recycling, 188.*

Hedemann, J. et al. (2022). *Kartierung des anthropogenen Lagers in Deutschland. Entwicklung eines dynamischen Stoffstrommodells und Aufbau einer Datenbank zur Prognose des Sekundärrohstoffaufkommens (KartAL II)*. Final report, 58.

Heidelberg.de. (2022). *Heidelberg wird kreislauffähig*. Accessed on 23.3.2023. https://www.heidelberg.de/hd/HD/service/23_06_2022+bergbau+in+der+stadt_+heidelberg+wird+europas+erste+kreislauffaehige+kommune.html

Heisel, F., & Hebel, D. (Ed.) (2021). *Urban mining und kreislaufgerechtes Bauen.*

Heller, N., & Flamme, S. (2020). Waste management of deconstructed External Thermal Insulation Composite Systems with expanded polystyrene in the future. *Waste Management & Research, 38*(4), 400–407.

Helmers, E. et al. (2022). Carbon footprinting of universities worldwide part ii: First quantification of complete embodied impacts of two campuses in Germany and Singapore. *Sustainability, 14.*

Hiebel, M., & Nühlen, J. (2016). Technical, economic, ecological and societal factors of steel scrap (Future Steel Scrap). In Technische, ökonomische, ökologische und gesellschaftliche Faktoren von Stahlschrott (Zukunft Stahlschrott). In Fraunhofer-Institut für Umwelt-, Sicherheits-und Energietechnik UMSICHT (Ed.), *Online version of the study commissioned by the Federal Association of German Steel Recycling and Disposal Companies eV (BDSV).*

Hillebrand, A. (2021). Kreisläufe schließen. In F. Heisel, & D. E. Hebel (Eds.), *Urban mining und kreislaufgerechtes Bauen—Die Stadt als Rohstofflage* (pp. 49–64).

Hobson, K. (2020). Small stories of closing loops': Social circularity and the everyday circular economy. *Climatic Change, 163*(1), 99–116.

HS Munich. (2021). *Press release: Kalkablagernde Bakterien—Aus Staub wird Gestein.* Accessed on 3.4.2022. https://www.presseportal.de/pm/143187/4823938

IRP. (2020). *Resource efficiency and climate change: material efficiency strategies for a low-carbon future. A report of the international resource panel.* United Nations Environment Programme.

Inside Climate News. (2022). *Concrete is worse for the climate than flying. why aren't more people talking about it?* https://insideclimatenews.org/todaysclimate/concrete-is-worse-for-the-climate-than-flying-why-arent-more-people-talking-about-it/

Jackson, M. D. et al. (2017), Phillipsite and Al-tobermorite mineral cements produced through low-temperature water-rock reactions in Roman marine concrete, *American Mineralogist: Journal of Earth and Planetary Materials,* 102(7), 1435–1450.

Jonas, H. (1979). *Das Prinzip Verantwortung.*

Kiefhaber. P. (2012). *Zukunftsentwicklung "Urban Mining"—Bauwerke als Rohstoffminen.* Lecture manuscript.

Kennedy, C., et al. (2011). The study of urban metabolism and its applications to urban planning and design. *Environmental Pollution, 159*(8–9), 1965–1973.

Kreft, O. (2016). Closed-loop recycling of autoclaved aerated concrete, geschlossener recyclingkreislauf für porenbeton. *Mauerwerk, 20*(3), 183–190.

Kreft, O. (2017). Sulfathaltiger Porenbeton: Ein umweltfreundlicher, langlebiger und recyclingfähiger Baustoff. *Mauerwerk, 21*(5), 287–296.

Kübler, S. et al. (2021). *Erarbeitung eines Rücknahmesystems am Beispiel des Cradle to Cradle® zertifizierten Würth Varifix® Schnellmontagesystems.* Nachhaltiger Konsum. Best Practices aus Wissenschaft, Unternehmenspraxis, Gesellschaft, Verwaltung und Politik, pp. 443–454.

Kurth, J. (2023). Closing the loop: Möbelrücknahme und Kreislaufwirtschaft, Vortrag gehalten auf dem 1. In *Internationales Wood BioEconomy-Forum (WBE).* Rosenheim University of Applied Sciences.

KWB. (2021). *Kreislaufwirtschaftsbericht bau.* https://kreislaufwirtschaft-bau.de/Download/Bericht-12.pdf

LGBR. (2019). *Reichweite von Vorräten nach Rohstoffgruppen.* https://lgrbwissen.lgrb-bw.de/rohstoffgeologie/rohstoffnutzung/rohstoffsicherung/reichweite-vorraeten-nach-rohstoffgruppen

Luhar, S., et al. (2020). Sustainable and renewable bio-based natural fibres and its application for 3D printed concrete: A review. *Sustainability, 12,* 24.

Mäbert, M., & Krug, D. (2023). *Neue Wege/Möglichkeiten (Recycling) der MDF-produktionlecture given at the 1st international Wood BioEconomy Forum (WBE).* Rosenheim University of Applied Sciences.

MAN. (2021). *First, large-scale CO_2-capture plant in cement production halves emissions.* Press release. https://www.man-es.com/company/press-releases/press-details/2021/12/17/first-large-scale-co2-capture-plant-in-cement-production-halves-emissions

McDonough, W., & Braungart, M. (2002). *Cradle to Cradle.*

Mettke, A. et al. (201). Erste Schritte zum urban mining. In W. Leal Filho (Ed.), *Aktuelle Ansätze zur Umsetzung der UN-Nachhaltigkeitsziele* (pp. 113–133).

Mettke, A. et al. (2019). Erste Schritte zum Urban Mining. In W. Leal Filho (Hrsg.), *Aktuelle Ansätze zur Umsetzung der UN-Nachhaltigkeitsziele.* 113–133.

Moholy-Nagy, S. (1968). *Die Stadt als Schicksal: Geschichte der urbanen welt.*

Moreno, M. et al. (2016). A conceptual framework for circular design. *Sustainability, 8*(9), 937ff.

Mostert, C. et al. (2021). Climate and resource footprint assessment and visualization of recycled concrete for circular economy. *Resources, Conservation and Recycling, 174.*

Müller, A., & Martins, I. (2022). *Recycling of building materials: Generation–processing–utilization.*

Müller, A., et al. (2013). Rückgewinnung sortenreiner Baustofffraktionen aus Mauerwerk. *Mauerwerk, 17*(6), 357–364.

Müller, A. (2012). Das sulfatproblem, Gipsabfälle. *Recycling Magazin, 22.*

Müller, A. (2018). *Baustoffrecycling: Entstehung-Aufbereitung-Verwertung.*

Nachtigall, W. (1997). *Vorbild Natur: Bionik-Design für funktionelles Gestalten.* Springer.

Neukirchen, F., & Ries, G. (2016). *Die Welt der Rohstoffe, Lagerstätten, Förderung und wirtschaftliche Aspekte.*

Nova Institute. (2021). *Global carbon demand for chemicals and derived materials.* Accessed on 30 Jan 2023. https://renewable-carbon.eu/publications/product/global-carbon-demand-for-chemicals-and-derived-materials-png/

Nußholz, J., et al. (2019). Circular building materials: Carbon saving potential and the role of business model innovation and public policy. *Resources, Conservation and Recycling, 141,* 308–316.

NZZ (2010). *Palast-Relikte von Ajaccio bis Paris.* https://www.nzz.ch/palast-relikte_von_ajaccio_bis_paris-ld.1265404. Retrieved: 13. Mai 2022.

Pietzsch, J. (2017). *Bioökonomie für Einsteiger.*

Prakash et al. (2022). *Modell deutschland circular economy.* Machbarkeitsstudie im Auftrag des WWF DeutschlandFreiburg.

Rees, W. (2018). *Ecological footprint* (pp. 43–48). Companion to environmental studies.

Ricker, M., et al. (2022). Statistical characterisation of reinforcement properties for textile-reinforced concrete: A novel approach. *Acta Polytechnica CTU Proceedings, 36,* 175–184.

Riechert, C. (2022). *Hydratation und Eigenschaften von Gips-Zement-Puzzolan-Bindemitteln mit alumosilikatischen Puzzolanen, Dissertation Uni Weimar.*

Risse, M. & Richter, K. (2018). *CaReWood-Cascading Recovered Wood, Teilvorhaben: Ökologische und ökonomische Bewertung der kaskadischen Holznutzung. Final report on the BMBEL research project.* TU München.

Rosen, A. (2021). Urban mining index—Planungs- und Bewertungsinstrument für zirkuläres Bauen. *Bauphysik, 43*(6), 357–365.

Saito, K. (2016). *Natur gegen Kapital: Marx' Ökologie in seiner unvollendeten Kritik des Kapitalismus.*

Sameer, H., et al. (2022). Comparative analysis of resource and climate footprints for different heating systems in building information modeling. *Buildings, 12,* 11.

Sarkodie, S. A. (2021). Environmental performance, biocapacity, carbon & ecological footprint of nations: Drivers, trends and mitigation options. *Science of the Total Environment, 751.*

Schneider, U. et al. (2022). Eine lebenswerte Stadt ist eine klimagerechte Stadt. Stadtplanung für menschliche Bedürfnisse ist klimagerecht. In J. Fritz (Ed.), *Transformationsgesellschaft: Visionen und Strategien für den sozialökologischen WandelMünster* (pp. 165–176).

Scope, C., et al. (2022). Life cycle assessment of carbon concrete composites: A circular economy path beyond climate mitigation? *A Cta Polytechnica CTU Proceedings, 33,* 518–524.

Sinnenwandel. (2016). *Hausbau nach C2C.* Accessed on 12.4.2022. www.sinnen-wandel.de

Sobek, W. (2023). Multilog mit Werner Sobek. Dem Ingenieur, Architekten und Vordenker des nachhaltigen Bauens zum 70. Geburtstag. *Nbau. Nachhaltig Bauen, 2*(2), 7.

Sojobi, A. O. et al. (2016). Recycling of polyethylene terephthalate (PET) plastic bottle wastes in bituminous asphaltic concrete. *Cogent Engineering.*

SRI. (2022). *Planetary boundaries.* Accessed on 12.1.2023. https://www.stockholmresilience.org/research/planetary-boundaries.html

Strunge, T. (2021). The costs of CO_2 carbonation in the cement industry. In N. A. Røkke, & H. K. Knuutila (Eds.), *TCCS–11CO2 Capture, transport and storage, short papers from the 11th international Trondheim CCS conference* (pp. 410–414).

Teischiner, A. (2023). *Spannungsfeld altholz und frischholzvorräte für energie und produkte?* Lecture given at the 1st International Wood BioEconomy Forum (WBE), Technical University Rosenheim.

Tercero Espinoza, L. et al. (2020). *The promise and limits of Urban Mining, Karlsruhe.*

Türk, O. (2014). *Stoffliche Nutzung nachwachsender Rohstoffe.* Springer.

Unnerstall, T. (2021). *Faktencheck Nachhaltigkeit.* Springer.

Weidner, S. et al. (2021). Graue Emissionen im Bauwesen-Bestandsaufnahme und Optimierungsstrategien. *Beton-und Stahlbetonbau,* 116(12), 969–977.

Weidner, S. et al. (2022). Ressourcenminimierung im urbanen Kontext. *Mauerwerk*, 26(4), 152–160.

Wörner, M. et al. (2016). Gradientenbetontechnologie: Von der Mischungsentwicklung über den Bauteilentwurf bis zur automatisierten Herstellung. *Beton- und Stahlbetonbau,* 111(12), 794–805.

Wuppertal Institut. (2010). *Website.* https://wupperinst.org/a/wi/a/s/ad/1129

Yang, X. et al. (2022). Urban mining potential to reduce primary material use and carbon emissions in the Dutch residential building sector. *Resources, Conservation and Recycling, 180.*

Youssef, H. et al. (2019). Dry laying of masonry to build demountable, highly energy-efficient prototype houses – First findings from a currently running research project. *Mauerwerk,* 23(4), 265–275.

Circular Bioeconomy of Buildings: Holistic Perspectives

7

Abstract

In the construction industry, a circular bioeconomy centers on rethinking the full lifecycle of materials, from planning through to the end of a building's use. This approach prioritizes conserving resources at every stage: reducing reliance on new raw materials, extending building lifespans, and ensuring that materials and components can be reused with high quality. At its core, this is about resource conservation. Renewable materials are gaining ground, supported by life cycle assessments focused on climate impact and bolstered by expanding technical possibilities. Nature's biological cycles offer inspiration for how to manage resources responsibly. Beyond the ideal of resource conserving planning and optimized value chains, bioeconomy is a matter of design an therefore understanding / human perception of building materials or architecture. This is addressed by understanding bioeconomy as a metaphor. This chapter outlines the key aspects of circular bioeconomy in construction, touching on everything from institutional frameworks and material flow management to technological innovations, material networks, and the cultural and communicative aspects that drive meaningful change.

Before sustainable construction can happen, a fundamental human condition should be met: an individual must hold a personal vision of (motivation for) sustainability, rooted in real experiences. This could involve memory of materials like wood, brick, or exposed concrete; reminder of being in beautifully experienced, diverse ecosystems; in vibrant or 'porous' cities (as Richard Sennett would say; Sennett et al., 2018); or witnessing the beauty of time-worn patinas in gnarled, aging forms. Such a vision is nor only rational, but sensually meaningful, stemming from a unique, first-person perspective, whether that of an architect, planner, or builder.

J. Grossarth, *Bioeconomy of Buildings*, https://doi.org/10.1007/978-3-031-84014-2_7

Fig. 7.1 A load-bearing pillar made of solid round timber integrates into a concrete and steel structure that has existed for around 70 years before. The sense of life conveyed by the architecture suggests a 'cultural interpretation' of bioeconomy. An interior view from the 'Weinlager' building in Basel by Esch Sintzel Architekten for Habitat Foundation. Photo Jan Grossarth

Ecosystems are shaped by the *meaning* we—or even animals—attribute to our surroundings (as theorized by Jakob von Uexküll). A mouse is meaningful for a cat, a house in its specific individual construction for us. Meaning arises exclusively from first-person perception. In the planning process, this meaning must be distilled into a conceptual framework of sustainability or bioeconomy. Yet, realizing this in practice often faces hurdles: economic trade-offs, planning and approval barriers, and the inherent dilemmas and goal conflicts of sustainability. Such a vision inspires deep engagement with materials, sparks innovation, and sometimes leads to the rediscovery of historic building techniques. Experience of meaning leads to deep interest. In this sense, sustainability is a gateway to creativity, not a corset that restricts creative freedom. The journey of planners and creators—their vision, struggles, and accomplishments— can be told as a narrative unto itself, one that embodies the depth and complexity of sustainable construction. This journey begins with a motivational impulse, takes through an enlightening and demanding process over months or years, and ends in a result: the building (Fig. 7.1).

More formally defined, the circular bioeconomy can be defined for the construction industry as follows:

> The circular bioeconomy in construction involves aligning planning and implementation with the lifecycle of the materials used. This includes resource-conserving raw material extraction, maximizing the longevity of buildings, and ensuring the high-quality reuse of

building materials and components. Resource conservation is the central concern. Renewable building materials and their cascading use are becoming increasingly important in climate balancing, thanks to technically enabled applications. Biological cycles can serve as an inspiration for sustainable resource management.

While the definition above is on the factual level, the following can be added regarding semantics:

The metaphor of the bioeconomy is discursively original. 'Bios,' meaning life, broadens the perspective to include the qualities of long-lasting architecture and the connection between quality of life and building culture. 'Oikos,' meaning household, grounds the sustainability concepts of building based on what is feasible in the face of financial, energetic, and material scarcities.

Reading the bioeconomy as a metaphor allows us to connect life semantics with industry logics and technical solution approaches. Understood in this way, this approach also makes building more of a public concern again.

There are many practical examples and innovative approaches for a smarter, more gentle use of building materials. This volume reflected an intense effort to provide a broad overview. The application space of biogenic materials is expanding due to significant process engineering and material science innovations. They are a nucleus of the *circular bioeconomy of buildings.*

As a applied science, it can collect role models for the many actors from the practice of construction, provide guidance as a guiding metaphor, so that their planning, construction, innovation and investment activities are consistently aligned with the *goal of resource conservation.* This results in emission reductions. Traditional ideas and metaphorical concepts like *cradle to cradle* set the corridors.

The borrowed science of the *circular bioeconomy of buildings* leans on much more that can be "inspiration" for institutional change: historical precedents, entrepreneurial case studies, biotechnological technology utopias, and *nature itself* as a model for technology and material flow management. Ethical considerations of land use or biotechnology applications follow. The *circular bioeconomy of buildings* also acts as a peace bridge in politics. As such, it enables novel conversations and collaborations—for example, among actors in construction with the goal of carbon neutrality. And it encourages to align funding and research policy with the goal of resource conservation where it is not already the case.

Looking at this goal of resource conservation and the conditions of its realisation or institutionalization—legally, economically, technologically, sociologically—one looks at the *factual level.* Looking at the sociocultural conditions of resource conservation—psychologically, ethnographically, historically—one looks at the *semantic level.* It is important for an understanding of acceptance, motivation, the anthropological reasons for the project, and thus practical relevance.

Essential fields of action and analysis of the *circular bioeconomy in construction* are summarised in the following twelve points.

7.1 Institutional Considerations

The construction industry is highly influenced by standards, by functional and industry standards. It is also characterized by multi-tiered value chains and competitive markets. This leads to high price and cost pressure at many levels (Chap. 2). Conservatism can be observed. Some speak of a "cementation" of the construction industry in self-made standards. However, these institutional rigidities are increasingly confronted with institutional change: changes in government requirements in the approval of substitute building materials (substitute building materials regulation), in the tendering practice in favor of secondary building materials, new take-back obligations or new standards that regulate the circularity of building products (such as demountability classes). But these often come with cost or application reservations and do not do justice to the dimension of the project of "transformation". The European *Green Deal* promises an epochal institutional change of the economic system from the 2020s onwards with the climate protection laws (Sect. 2.2), but the acceptance in the population and thus the conditions of realisation are still unclear. The governmental framework for a *circular bioeconomy* is prepared at the European level (Sect. 6.6). Thus, entrepreneurial and technical-scientific innovation activity is given a new orientation, which could have the qualities of a disruptiveness of *Schumpeterian* magnitude (Pyka, 2017). Also, the systematically increasing CO_2 certificate prices in Germany and the EU and the increasing ESG reporting obligations within the framework of the taxonomy to the life cycle approach (grey energy) are increasingly changing the economic calculus of the actor groups in the construction industry. However, standards may also have to be lowered if carbon neutrality is seriously considered as a priority sustainability goal—such as standard requirements for fire protection, sound insulation or maximum daily loads of roads.

7.2 Material Flow Management

Many countries improved their *recycling rates* also in construction, what can be seen as a remarkably successful application practice of the recycling and reuse technologies of construction waste (Sect. 6.2). In waste management law and in the Construction Products Regulation, a cascading material management is already laid out, first take-back obligations of the manufacturers are in force. Legal considerations must be weighed against the socio-economic interests of the building material manufacturers. Digital trading platforms are revitalizing the used building material trade. But the chemical contaminations of past decades still pose a long-term barrier to truly utilizing the "urban mine" according to the idea of reuse or upcycling (Sect. 7.5). Here, old wood is the prime example. The importance of *Closed-Loop materials* such as clay (bricks) is likely to increase significantly in view of the goal of circular construction, that of *Loop materials* such as clinker is likely to decrease relatively. Life cycle analysis will become a standard part of every bachelor's degree in construction professions.

7.3 Technical Innovations

Technical change can bring about behavior-neutral improvements in the Environmental consequences of consumption. An intensification of the circular flow of mineral and biogenic material flows is expected from various biotechnical innovations (Chap. 5). These include bacteria, insects or fungi (mycelium; Sect. 5.4) that metabolise organic biomass. Thus, the material flows of agricultural and organic waste biomass and the construction industry are increasingly interlocking. *Genome Editing* can play a role in increasing resource efficiency and thus the economic attractiveness of establishing these processes. The "plastic" producing bacterium *C Necator* is an example, as only a genome change significantly increased the conversion efficiency from feed fats to "bioplastic" and brought this technology close to market readiness by the beginning of the 2020s (Sect. 4.3). Environmental technology works on many levels: irrigation hoses for economical irrigation are one example, desalination plants another. Adjustments of lignin contents of poplars is an example that directly affects the construction industry (Sect. 4.10). Sensor-supported separation technologies of construction waste is another that allows for more pure separation (Sect. 6.2.5). Gradient concrete or CCS technologies that can make cement production significantly more climate-friendly are further examples (Sect. 6.3). However, these only become "game changers" when the cost–benefit ratio of research, development and establishment in industrial production or processing enables market readiness.

7.4 Integrated Perspectives on Agriculture, Forestry and Construction

Overall, the networking of the agricultural and forestry sectors with the construction industry is significantly increased by the *circular bioeconomy*. Also, additive manufacturing using lignocellulose-based raw materials (3D printing of "wood") increases the demand for "biomass" from field, forest and pasture (Sect. 3.7). For waste, especially old wood and lignocellulose—for example from agricultural secondary material flows—increasing recycling options in the sense of "upcycling" are opening up. The broad horizon of agriculture requires practitioners in the construction industry, at least in high management, to have knowledge and skills to justify the problem of land competition and ecology when a company relies on such materials. One could call this "bioeconomic language skills"—or in this sense: *ethical reflection ability*. Ecologisations and climate adaptations of agriculture such as agroforestry systems and landscape structure elements (hedges, rows of trees) also indirectly provide new raw material sources for construction. In agriculture, which traditionally sees itself as the nation's nourisher, there is still a need for information. Straw as a by-product of grain production could have a great career as a construction material, because its availability in Europe is more than sufficient and its talents as a carbon store are excellent. Hemp and cork are less mass relevant.

7.5 Urban Mining

The global urban mine is estimated to hold more material that total dry biomass weights (Fig. 7.2). Construction in existing buildings, selective demolition and a pure separation using innovative separation techniques—such as a sensory separation by colors—are expected to gain in importance. The materials obtained in this way can improve the climate-related life cycle assessments on the raw material side. While the method of building life cycle assessment allows the reuse and recycling options of components after the end of a building life cycle only beyond the system boundary—and/or in the life cycle assessment of the following building—such considerations in the *circular design* of buildings fall into specialized indices like the *Urban Mining Index* carries more weight. The European *Madaster* Circularity Index does something similar. The private sector *C2C* building certification also places value on circularity. Here, a particularly attractive narrative about the positive contributions of the construction industry to environmental *improvement* is achieved—instead of aiming at politically popular semantics of "climate guilt" or "planetary boundaries". These approaches are related to a supplement

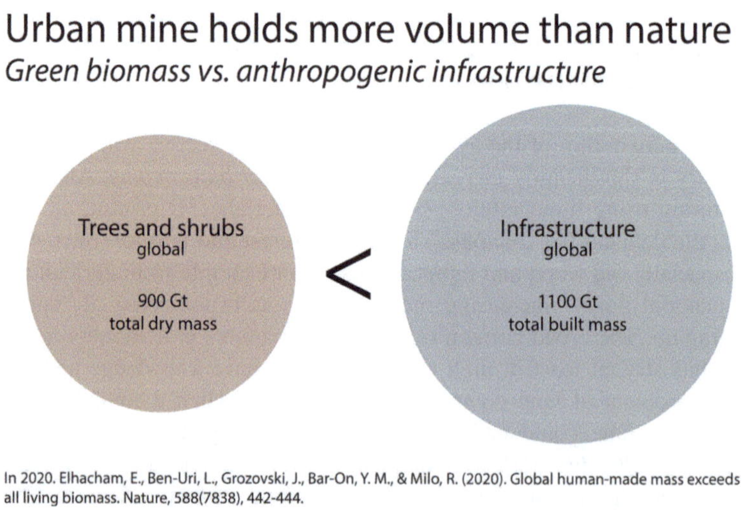

Fig. 7.2 Urban mine holds more material weight than dry biomass. JG; Elhacham et al. (2020)

to the "old" energy efficiency approach to the life cycle approach—and beyond. Through technical or product innovations, worthless materials can be made usable. The re- and *upcycling* of PET bottles, which can even become building blocks for houses, or the use of desert sand as filler for new types of building blocks are just two examples.

7.6 Project Communication

The many, sometimes competing, interests in the concrete design of a 'green transformation' (including some resistance to the project) require dialogue and moderation. Political inertia should not only be understood by referring to sociological models of modern society but also taken seriously and respected as a phenomenon of modern democratic societies. Early project communication and moderation in the concrete construction project—both internally and externally—from the pre-planning phase increase the chances of achieving sustainability orientation. There appears to be a momentum of inertia in construction planning. New forms of cooperation between various actors and stakeholders are being tested in several European cities, such as Malmö and Heidelberg. These approaches are scientifically recommended. Compared to the challenges of other sectors (aviation, energy sector, food systems), the material-saving, circular construction is likely to present a less gigantic technical challenge. Building products based on the "Lego principle" (the project *ReMoMab* is just one example, Sect. 5.2.6) are technically mature. In fact, they apparently fail due to the low economic motivation of the construction industry to bring these "into the field". But actually, there seems to be an awareness of the need for a transformation towards resource conservation among the actors in the construction industry, as the expert interviews for this book clearly suggest, some of which even see a pressure for change from "external forces".

7.7 Early Awareness of Resource Scarcities

To make an awareness and the effort to prevent resource shortages at an early stage a *truly action-relevant* motive, the construction professional world would first need more factual clarity in this field. However, even science does not have this. In fact, it is scientifically difficult to determine how far the reserves of mass construction materials extend. We often read about sand shortages in media examples, but science is vague here. Projections change again and again. There are historical experiences with false predictions (*Peak Oil*). All of this carries the risk that the dangers of scarcity will be ignored. However, there are official forecasts of a shortage of gypsum, and also of regional shortages of sandy gravels, carbonate rocks and limestones in southern Germany. The metal content of newly mined ores has already dropped significantly in some cases, as the example of copper shows (see preface). The raw material side is more important in the long term than reducing emissions. Because if the usable mineral and metallic raw materials as an industrial basis become scarce, a society—figuratively speaking—must "burn everything", as the high usage rate of wood for cooking and heating in many countries of the African continent shows.

7.8 Bioeconomic Thinking as De-ideologisation

The approach of the *circular bioeconomy of buildings* includes not only technical and organisational, but also political values of hope. It is suitable for de-ideologizing the project of transformation. Ideas of sustainability and the *economic as a level of action* are not dialectically played off against each other in this metaphor, but originally connected with each other.

It encourages "dialectical" thinking in the awareness of conflicting goals but also for concretisation through local observation. The *historical review* of the ideas, agricultural and energy history as part of the bioeconomy makes a further contribution to classification and shifts the focus from the scientifically highly justified, media-resonant, but often little practical *climate policy disaster mode* to historical-technological coping achievements, but also disaster events. It urges decisive action and reinforces the hope in human problem-solving ability.

The "Malthusian" seeming prophecy of doom of the "founder" of the *political-economic bioeconomics*, Georgescu-Roegen, is understandable in the background of his life and time history and his outsider role in a narrow-minded academic world comprehensible. And his writings are also fruitful for the transdisciplinary further development of the bioeconomic discipline, to which his original wealth of images and metaphors contributes, with which he outlines the visions of the "entropic" catastrophe. Georgescu-Roegen's almost poetic language tradition of bioeconomics testifies to the *connection of experience, observation, abstraction and formalism*, which does justice to the wide horizon that the metaphor opens.

The "agricultural age" so intensively reviewed and traveled by Georgescu-Roegen and biographically experienced is from the perspective of the (construction) bioeconomy as a *negative foil of relapse* to consider. The memory of the agricultural is like a reminder that we as modern humanity cannot survive without techniques, have become dependent on technical development. The distant mirror of the agricultural is therefore also *fruitful* to use as a positive image reservoir and as a source of inspiration for the rediscovery of a characterful as well as energy-saving construction. One could almost say: a *simple construction.*

For all these reasons, this bioeconomic perspective "liberates" from having to discuss sustainability along *binary categories*: "Wood or concrete?" "Electric motor or bioethanol?" "Meat or soy?".

The added value of semantic sensitivity can be illustrated using the last question: *Vegetarian* by definition—or: on the factual level—means categorically meatless. Vegetarian, from the etymology, also means *invigorating*, alive (lat. *vegetabilis*). In this sense, one could ask: To what extent could meat play a role in a diet that we speak of as "invigorating"? What role does the origin, the growth and the life story of an animal? Is it really about the 'thing,' or about relationships? About life, about the stories of building materials? Wouldn't such a semantically reflected standpoint of judgment actually be more motivating to reduce the proportion of climate-damaging mass livestock meat than a literally understood vegetarian diet? Especially when this reduction is motivated by considerations of guilt? Don't creativity and material, as well as historical interconnectedness with things, go hand in hand? In this sense, from the perspective of a broadly conceived bioeconomy, the historicity of buildings would also matter.

7.9 Transdisciplinary and Hermeneutic Research

Since the internationally significant, *resource-oriented tradition* of bioeconomy indeed runs the risk of being overbearing towards the living in view of the utility value of "biomass", a look at *qualities* of a bioeconomic "transformation" of construction is absolutely necessary. For the scientific accompaniment, a problem-oriented and transdisciplinary research was therefore expressly recommended, which does not exclude ethics, ethnography and semiotics. Non-scientific, societal actors are included by taking into account their impression of buildings. Hermeneutics becomes part of the technology assessment.

7.10 Historicity in Building

The unsustainability of construction in terms of its high consumption of resources can be described in various ways. This book is rich in examples. The waste statistics tell of it, the results of the life cycle assessments or lifetimes of buildings. In addition, there is the thesis that buildings have lost their uniqueness, character, individuality compared to earlier times. Often the problem is lamented as "investor architecture" or with other metaphors. The *culture of clearance* of the 1950s—rapid demolition, disposal, new construction—has radically changed the practices of deconstruction and new construction. The economisation of construction—even if it historically brought affordable housing and unprecedented large living spaces per capita, and thus served the goal of "social sustainability"—should also be perceived as a sustainability problem. The conceptual bridge into the "language" of life cycle assessment is the fatal *short-livedness* of those buildings that construction has produced in the past approximately 60 years. It is historically unprecedented. In this regard, in this introductory book not only the concepts of *effect, atmosphere, beauty* and *historicity* were introduced, but also related to the term *building culture* (Sect. 2.3). A project by two architecture graduates from the University of Weimar to document and digitally record the "salvage stories" of building materials that are reused, fits very well with these thoughts (Iannone and Heyer 2023). The sociocultural background events are a decisive key to understanding the complexity of the sustainability crisis. Individualisation, lifestyle society, consumer society are suitable as theoretical explanations to explain the loss of the claim of the builders to contribute to the public space with the construction (Sect. 2.3.1.5).

7.11 Passion and Craftsmanship in Construction

The approach of the bioeconomy forces academic approaches to sustainability to "converse" with the lifeworld. Craftsmanship and Passion for construction plays a significant role in the transmission of “head” and “hand”. As a testament to artisanal passion, this monograph recalls the words of Wendell Berry—namely his “historical” speech about an agrarian culture. The pride in the house, the craft, the *fruits of labor* seemed to him crucial for the *good life* in general. Industrialisation, individualisation, and also technical academisation play ambivalent roles in this regard. In 1974, Berry said of industrialised agriculture: “And it’s a work of monsters’ ignorance and irresponsibility on the part of the experts and politicians.” This was not meant to be technophobic, but out of justified concern that people, in the face of technical power and in an increasingly abstract relationship to the object of their contemplation and creation, are no longer able to perceive that people want to live and work in a community-oriented, history-oriented, experience-rich, value-bound, and hopeful manner. Georgescu-Roegen spoke of “exosomatic power” that man had gained through technology. Wendell Berry’s thoughts are compatible with newer perspectives from neuro-, psychological, and psychiatric sciences, as well as phenomenology, which under the umbrella term *embodiment* bring the bodily sensation back into focus—for example, of harmonious architecture. The bodily sensation precedes rational judgement. Why do people love the buildings of the founding period so much? How are awe, sympathy, and longevity of buildings related?

7.12 Bioeconomy as a Metaphor

The metaphor of the bioeconomy connects memory, experience, and physicality with the factual level: building materials and their engineering characteristics, material flows in the economy, resource availability, and life cycle balances. It allows emotion, sensation, and personality to be articulated indirectly where strict objectification norms prevail—in (specialist) politics, at investor conferences, and at civil engineering conferences. In the word part ‘Bios,’ life, the meanings of everyday life, sense of life, dignity and joy, pain, and failure are hidden, albeit as inaccessible as ‘index fossils’ (Hans Blumenberg). Though it may feel somewhat awkward to express this in an engineering science publication, the original contribution of this book chapter is to link the ‘depth’ of questions about good and simple construction with the bioeconomy discourse. One sense of academic discussion and research about sustainability is that it provides a way out of the sometimes technocratic prisons of professional worlds and specialist towers, guiding practitioners and entrepreneurs in the construction industry back to the dignity of making decisions rather than simply meeting the needs of residential and real estate consumers. Quite simply. Perhaps, despite all the merits of this ancient concept, we should talk less about sustainability and more about the successful and meaningful effort and dialogue around creating, converting, and deconstructing in ways that are meaningful, beautiful, and enjoyable—for people of this and future generations, and for the animal and plant world. The metaphor of the bioeconomy supports this view and thus goes beyond the scientifically influenced concept of bioeconomy. This means, in a way, supplementing the imperative of planetary boundaries with the cultivation of planetary sensitivity. Metaphors are not to be taken literally, but seriously.

7.13 Conclusion

Building is also a form of communication. Communication has both a factual level and a level of meaning. When construction professionals such as building material manufacturers and engineers do not (or can not) focus on this in their professional action—partly due to the unfortunate professional split with architecture—it is not only their problem but also that of future generations.

Academic approaches to the bioeconomy initially complicate efforts for sustainability. However, they also offer numerous approaches for implementation in practice. Practically, it makes sense to simplify from complexity and decide on an approach. If, in the current efforts for sustainability, 'simple building' finds great resonance, it is also to be understood that the good wants to be recognized 'simply.' The sense for sustainable building joins the numerous norms and partly legally defined standards, such as the ESG criteria, life cycle assessment data, and the resource and land conflict theses, which ethically always have to be considered.

The roof of the bioeconomy extends in many ways over the factual and meaning levels of resource conservation. This cannot be said of the decades-long discussion of sustainability. Bioeconomic thinking does not invite us to separate the factual and relational levels from each other or to play them off against each other. In this lies the chance to prevent, in the future, what Michael Braungart correctly says in the preface: 'Our sustainability expertocracy helps to optimize the wrong. Bioeconomy as a metaphor also does not block the sensual reason of sustainability. This was already laid out in the writings of its founder Georgescu-Roegen (whatever one may think of his civilisational pessimism).

Finally, a last extension. A metaphorical understanding of the concerns of the environmental movement is excellently expressed in the internationally much and also far beyond church circles noticed so-called "eco-encyclical" *Laudato Si* of Pope Francis from the year 2015. About the climate crisis it says therein:

> If we approach nature and the environment without this openness to awe and wonder, if we no longer speak the language of fraternity and beauty in our relationship with the world, our attitude will be that of masters, consumers, ruthless exploiters, unable to set limits on their immediate needs. By contrast, if we feel intimately united with all that exists, then sobriety and care will well up spontaneously. The poverty and austerity of Saint Francis were no

> mere veneer of asceticism, but something much more radical: a refusal to turn reality into an object simply to be used and controlled (Francis, 2015, p. 11).

In this respect, bioeconomy can invite to a special attitude that is beyond ideological camps, also to a language of vitality in connection with the acceptance of needs of technical innovation. Its vocabulary does not exclude terms like: awe, relationship, wonder, beauty, care and—simplicity. Sufficiency is not an end in itself.

This book (and this chapter) was primarily intended to compile scientific facts and analyze their practical relevance, also against the background of the multi-layered discursivity of sustainability concepts. However, it was also concerned with considering the anthropological dimension of resource conservation. Awe is certainly not the core concern of civil engineering science and other technical disciplines, to which modern humanity owes so much quality of life. But the construction industry can also develop in view of the ecological crises 'along' lifeworld phenomena. Buzzwords from the academic ivory tower, such as 'transdisciplinary,' might not be very inviting for this. More important might be to adequately understand the ecological crisis and one's own entanglement in its causes. Knowledge in sustainability issues can lead to more reflective action by technical professionals. The ability to marvel precedes the imagination of attractive technological futures. This is how bioeconomic construction innovation arises. The technical prerequisites are created in the painstaking detail work of the laboratories.

References

Elhacham, E., et al. (2020). Global human-made mass exceeds all living biomass. *Nature, 588*(7838), 442–444.

Francis. (2015). Laudato si: On care for our common home.

Iannone, N., & Heyer, M. (2023). Materialgeschichten. *nbau. Nachhaltig Bauen, 2*(2), 59–60.

Pyka, A. (2017). Die Bioökonomie unter dem Blickwinkel der Innovationsökonomie. In Pietzsch, J. (Ed.), *Bioökonomie für Einsteiger* (pp. 129–138).

Sennett, R. et al. (2018). The science of urbanization and the open city. In *The Quito papers and the new urban agenda* (pp. 52–149).

MIX
Papier aus verantwortungsvollen Quellen
Paper from responsible sources
FSC® C105338

If you have any concerns about our products,
you can contact us on
ProductSafety@springernature.com

In case Publisher is established outside the EU,
the EU authorized representative is:
Springer Nature Customer Service Center GmbH
Europaplatz 3, 69115 Heidelberg, Germany

Printed by Libri Plureos GmbH
in Hamburg, Germany